BRITISH RAILWAYS

COAC

STOCK

EIGHTEENTH EDITION – 1994

The Complete Guide to all
BR Loco-hauled Coaching Stock
(excluding departmental stock)

Peter Fox & Richard Bolsover

ISBN 1 872524 58 3

© 1993. Platform 5 Publishing Ltd., Wyvern House, Sark Road, Sheffield, S2 4HG.

CONTENTS

INTRODUCTION

This book, formerly known as 'Coaching Stock Pocket Book', contains full details and differences of all BR locomotive-hauled coaching stock. We would like to thank all who have helped by contributing information or photographs. Information is updated to 8th November 1993.

Coaches are listed in batches, according to their class, with lot number information for the various batches being shown above the listings. Where a coach has been renumbered, the former number is shown in parentheses. If the coach has been renumbered more than once, the original number is shown in parentheses, with the intermediate numbers being given in the text. Where the old number of a coach due to be converted or renumbered is known and the conversion or renumbering has not yet taken place, the coach is listed both under its old number with its depot allocation, and under its new number without an allocation

In this issue, coaches with an 'HQ' allocation, i.e a BR Headquarters responsibility, are listed as such and a separate section giving storage locations is provided.

Note: Owing to a keying-in error, the 1993 edition, which was the 17th edition, was inadvertently titled the 18th edition. We apologise for this error.

NUMBERING SYSTEMS

Six different numbering systems were in use on BR. These were the BR series, the four pre-nationalisation companies' series' and the Pullman Car Company's series. Only BR number series loco-hauled coaches now exist in stock.

All coaching stock vehicles have now been given depot allocations, regional prefixes not now being used.

DETAILED INFORMATION AND CODES

After the heading, the following details are shown:

(1) Diagram code. This consists of the first three characters of the TOPS code followed by two numbers which relate to the particular design of vehicle.
(2) 'Mark' of coach (see below).
(3) Number of first class seats , standard class seats and lavatory compartments shown as nF nS nL respectively.
(4) Bogie type (see below).
(5) Brake type. (see below).
(6) Heating type. (see below).
(7) Additional features.
(8) ETH Index.

BOGIE TYPES

BR Mk 1 (BR1). Standard double bolster leaf spring bogie. Generally 90 m.p.h. but certain vehicles were allowed to run at 100 m.p.h. with special maintenance. Weight: 6.1 t.
BR Mk 1 (heavy duty). Similar to above. Weight 6.5 t.
BR Mk 2 (BR2). Later variant of BR Mark 1 used on certain vans.
COMMONWEALTH (C). Heavy, cast steel soil spring bogie. 100 m.p.h. Weight: 6.75 t.
B4. Coil spring fabricated bogie for 100 m.p.h. Certain BGs (classified NHA) are allowed to run at 110 m.p.h. with special maintenance. Weight: 5.2 t.
B5. Heavy duty version of B4. 100 m.p.h. Weight: 5.3 t.
BT10. A fabricated bogie designed for 125 m.p.h. Air suspension.
T4. The latest 125 m.p.h. bogie from BREL.
The new Mark 4 vehicles are fitted with bogie from the Swiss firm of SIG.

BRAKE TYPE CODES.

a   Air braked.
v   Vacuum braked.
x   Dual braked (air and vacuum).

HEATING TYPE CODES.

d   Dual heated (steam & electric).
e   electric heated
o   No heating apparatus.

r   Steam piped & electric wired.
u   UIC/BR electric heat.
y   Electric wired.
Note: All BR train heating nowadays is electric, but dual heated vehicles still often retain the steam heating equipment, albeit isolated.

ADDITIONAL FEATURE CODES.

f   Facelifted or fluorescent lighting provided.
k   Composition brake blocks (instead of cast iron).
n   Day/night lighting.
p   Fitted with public telephone.
pa   Public address speakers installed.
pg   Public address transmission and driver – guard communication.
pt   Public address transmission and speakers.
q   Fitted with catering staff to shore telephone.
s   Short swing links (see page 7)
to   Public address transmission only.
w   Fitted with wheelchair space.
z   Fitted with wheelchair space and disabled persons' toilet.

NOTES ON ETH INDICES.

The sum of ETH indices in a train must not be more than that of the locomotive. Suffix 'S' was used to denote SR 750 V heaters, the usual voltage on BR being 1000, and suffix 'X' denotes 600 amp wiring instead of 400 amp. Trains whose ETH index comes to more than 66 must be formed completely with 600 amp wired stock. There are now no loco-hauled vehicles in BR capital stock with SR heaters.

TOPS CODES

TOPS (Total operations processing system) codes are allocated to all coaching stock. For passenger stock the code consists of:

(1) Two letters denoting the layout of the vehicle as follows:

AA Gangwayed Corridor
AB Gangwayed Corridor Brake
AC Gangwayed Open (2 + 2 seating)
AD Gangwayed Open (2 + 1 seating)
AE Gangwayed Open Brake
AF Gangwayed Driving Open Brake
AG Micro-Buffet
AH Brake Micro-Buffet
AI As 'AC' but fitted with drop-head buckeye and no gangway at one end.
AJ Restaurant Buffet with Kitchen
AK Kitchen Car
AL As 'AC' but with disabled person's toilet (Mark 4 only)
AN Miniature Buffet
AO Privately-owned
AS Sleeping Car
AT Royal Train Coach
AU Sleeping Car with Pantry

(2) A digit for the class of passenger accommodation:

1   first
2   standard (formerly second)
3   Composite
4   Unclassified
5   None

(3) A suffix relating to the build of coach.

1 Mark 1	A Mark 2A	C Mark 2C	E Mark 2E	G Mark 3 or	H Mark 3B
Z Mark 2	B Mark 2B	D Mark 2D	F Mark 2F	Mark 3A	J Mark 4

For non-passenger carrying coaching stock, the suffix denotes the brake type:

A   Air braked
V   Vacuum braked
X   Dual braked

OPERATOR CODES

The normal operator codes are given in brackets after the TOPS codes. These are as follows:

F   First
S   standard (formerly known as 'second')
C   Composite
B   Brake
O   Open
K   Side corridor with lavatory

Various other letters are in use and the meaning of these can be ascertained by referring to the titles at the head of each class.

ABBREVIATIONS:

DMU	Diesel multiple unit
GER	Great Eastern Railway
GWR	Great Western Railway
LMS	London Midland and Scottish Railway
LNER	London & North Eastern Railway
LNWR	London & North Western Railway

THE DEVELOPMENT OF BR STANDARD COACHES

The standard BR coach built from 1951 to 1963 is the mark 1. This has a separate underframe and body. The underframe is normally 64'6'' long, but certain vehicles were built on short (57') frames. Tungsten lighting is standard and until 1961, BR mark 1 bogies were generally provided. In 1959 TSOs to lot No. 30525 appeared with fluorescent lighting and melamine interior panels and from 1961 onwards Commonwealth bogies were fitted in an attempt to improve the quality of ride which became very poor when the tyre profiles on the wheels of the Mark 1 bogies became worn. The further batches of TSOs and BSOs retained the features of lot 30525, but the BSKs, SKs, BCKs and CKs, whilst utilising melamine panelling in standard class, still retained tungsten lighting. Wooden interior finish was retained in first class compartments. The FOs had fluorescent lighting with wooden panelling except for lot No. 30648 which had tungsten lighting. In later years many mark 1s had their mark 1 bogies replaced by B4s.

In 1964, a new train was introduced. Known as ''XP64'', it featured new seat designs, pressure ventilation, aluminium compartment doors and corridor partitions, foot pedal operated toilets, and B4 bogies. The vehicles were on standard mark 1 underframes. Folding doors were fitted but these proved troublesome and were later replaced with hinged doors. All XP64 coaches have now been withdrawn, but some have been preserved.

The prototype mark 2 vehicle (W 13252) was produced in 1963. This was an FK of semi-integral construction and was pressure ventilated. Tungsten lighting was provided and B4 bogies. This vehicle has been preserved by the National Railway Museum. The production build was similar, but wider windows were used. The standard class open vehicles used the new seat design similar to that in the XP64 and fluorescent lighting was provided. Interior finish reverted to wood. MK 2s were built from 1964 – 66.

The mark 2As, built 1967 – 68, incorporated the rest of the novel features first used in the XP64 set, i.e. foot pedal operated toilets (except BSOs), new first class seat design, aluminium compartment doors and partitions together with fluorescent lighting in first class compartments. Folding gangway doors (lime green coloured) were used instead of the traditional variety. The following list summarises the changes made in the later Mk 2 variants:

Mk 2B: Wide wrap round doors, no centre doors, slightly longer body. In standard class, one toilet at each end instead of two at one end as previously . Red gangway doors.
Mk 2C: Lowered ceiling with twin strips of fluorescent lighting, ducting for air conditioning, but no air conditioning.
Mk 2D: Air conditioning. No opening lights in windows.
Mk 2E: Smaller toilets with luggage racks opposite. Fawn gangway doors.
Mk 2F: Plastic interior panels. Inter-City 70 seats. Modified air conditioning system.

The Mark 3 coach has BT10 bogies, is 75' long and is of fully integral construction with Inter-City 70 seats. Gangway doors are yellow (red in RFB). Loco-hauled coaches are classified Mark 3A, Mark 3 being reserved for HST trailers. A new batch of FOs and BFOs classified Mark 3B was built in 1985 with APT style seating and revised lighting. The last vehicles in the Mark 3 series are the driving brake vehicles (officially called driving van trailers) which have been built for West Coast Main Line services.

The Mark 4 coach built by Metro-Cammell for the East Coast Main Line electrification scheme features a body profile suitable for tilting trains, although tilt is not fitted, and is not intended to be. They are suitable for 140 m.p.h. running, although initially they will be restricted to 125 m.p.h. pending the installation of automatic train protection (ATP) on the East Coast Main Line.

LOCO-HAULED PASSENGER STOCK

AJ11 (RF)   RESTAURANT FIRST

Dia. AJ106. Mark 1. Gas cooking. 24F. B5 bogies. ETH 2. This coach spent most of its life as a Royal train vehicle 2907.

Lot No. 30633 Swindon 1961. 41 t.

325   I   ICHV   BN

AJ1F(RFB)   BUFFET OPEN FIRST

Dia. AJ104. Mark 2F. Air conditioned. Converted 1988 – 9/91 at BREL, Derby from Mark 2F FOs. 1200/1/3/6/11/14 – 17/20/21/50/2/5/6/9 have Stones equipment, others have Temperature Ltd. 26F 1L. B4 bogies. ae. pt. payphone. Catering staff – shore telephone. ETH 6X.

1200/3/6/11/14/16/20/52/5/6. Lot No. 30845 Derby 1973. 33 t.
1201/4/5/7/8/10/12/13/15/17 – 9/21/50/1/4/7/9. Lot No. 30859 Derby 1973 – 4. 33 t.
1202/9/53/8. Lot No. 30873 Derby 1974 – 5. 33 t.

Note: 1200 – 2/5/7/9/10/2/4/7/20 were also numbered 6459/45/56/38/22/57/62/53/33/44/32 respectively when declassified prior to conversion.

1200 (3287)	I	ICCX	MA	1216 (3302)	I	ICCX	PC
1201 (3361)	I	ICCX	PC	1217 (3357)	I	ICCX	PC
1202 (3436)	I	ICCX	PC	1218 (3332)	I	ICCX	MA
1203 (3291)	I	ICCX	MA	1219 (3418)	I	ICCX	MA
1204 (3401)	I	ICCX	PC	1220 (3315)	I	ICCX	PC
1205 (3329)	I	ICCX	PC	1221 (3371)	I	ICCX	MA
1206 (3319)	I	ICCX	PC	1250 (3372)	I	ICCX	MA
1207 (3328)	I	ICCX	PC	1251 (3383)	I	ICCX	MA
1208 (3393)	I	ICCX	PC	1252 (3280)	I	ICCX	MA
1209 (3437)	I	ICCX	PC	1253 (3432)	I	ICCX	MA
1210 (3405)	I	ICCX	PC	1254 (3391)	I	ICCX	MA
1211 (3305)	I	ICCX	PC	1255 (3284)	I	ICCX	MA
1212 (3427)	I	ICCX	PC	1256 (3296)	I	ICCX	MA
1213 (3419)	I	ICCX	PC	1258 (3322)	I	ICCX	MA
1214 (3317)	I	ICCX	MA	1259 (3439)	I	ICCX	MA
1215 (3377)	I	ICCX	PC	1260 (3378)	I	ICCX	MA

AJ41 (RBR)   RESTAURANT BUFFET

Dia. AJ403. Mark 1. Gas cooking. Built with 23 loose chairs (dia. AJ402). All remaining vehicles refurbished with 23 (21 w) fixed polypropylene chairs and fluorescent lighting. Commonwealth bogies. pt. ETH 2 (2X*).

Coaches with suffix 'R' in sector code have been further refurbished. 21 chairs, payphone, wheelchair space and carpets (Dia. AJ417).

Note: 1680 is on loan to Flying Scotsman Railways (set BN91).

Lot No. 30628 Pressed Steel 1960 – 61. d. 39 t.

1644 a	I	ICHH	BN		1674 a	I	ICHV	BN
1645 a	I	ICHV	BN		1675 x*	I	ICHV	BN
1646 a	I	IXXZ	HQ		1678 x*	I	ICHV	BN
1647 a	I	IANR	NC		1679 a	I	ICHV	BN
1649 aw	I	IXXZ	HQ		1680 x*w	M	ICHL	BN
1650 aw	I	ICHH	BN		1683 a p	I	IANR	NC
1652 aw	I	ICHH	BN		1684 x*		ICHH	CL
1653 aw	I	ICHV	BN		1686 a p	I	IANR	NC
1655 a	I	ICHH	CL		1688 aw		ICHH	BN
1658 a	I	ICHV	BN		1689 a p	I	IANR	NC
1659 a	I	ICHV	BN		1691 a p	I	IANR	NC
1663 x*	I	ICHH	BN		1692 a p	I	IANR	NC
1666 x*	I	ICHH	BN		1693 x*	I	ICHV	BN
1667 x	I	ICHV	BN		1696 a p	I	ICHX	CL
1670 x*w		ICHV	BN		1697 a p	I	IANR	NC
1671 x* p	I	ICHX	BN		1698 a	I	ICHV	BN
1672 x*	I	ICHH	BN		1699 a p	I	IANR	NC
1673 aw		IXXZ	HQ					

AN21 (RMB)   MINIATURE BUFFET CAR

Dia. AN203. Mark 1. Gas cooking. 44S 2L. These vehicles are basically an open standard with two full window spaces removed to accommodate a buffet counter, and four seats removed to for a stock cupboard. All remaining vehicles now have fluorescent lighting. All vehicles have Commonwealth bogies except 1850 (B5). d. ETH 3.

1832. Lot No. 30520 Wolverton 1960. 38 t.
1842 – 1850. Lot No. 30507 Wolverton 1960. 37 t (1850 is 36 t).
1853. Lot No. 30670 Wolverton 1961 – 2. 38 t.
1871. Lot No. 30702 Wolverton 1962. 38 t.

1842/50/71 have been been refurbished and are fitted with a microwave oven and payphone. Dia. AN208.

1832 x	I	ICHD	CL		1853 x	I	ICHD	CL
1842 x pt	I	ICCR	MA		1871 x pt	I	ICCR	PC
1850 a pt	I	ICCR	DY					

AJ41 (RBR)   RESTAURANT BUFFET

Dia. AJ414. Mark 1. Gas cooking. These vehicles were built as unclassified restaurant (RU). All remaining vehicles were rebuilt with buffet counter and 21 fixed polypropylene chairs (RBS). They were then further refurbished by fitting fluorescent lighting and reclassified RBR. ad. w pt. ETH 2X.

Lot No. 30632 Ashford/Eastleigh 1960 – 61. Commonwealth bogies. 39 t.

1959	I	IXXZ	HQ		1972	I	IXXZ	HQ
1966	I	IXXZ	HQ		1984	I	IXXZ	HQ
1971	I	IXXZ	HQ					

AU51   CHARTER TRAIN STAFF COACHES

Dia. AU501. Mark 1. Converted from BCKs. ETH 2.

Lot No. 30732 Derby 1964. Commonwealth bogies. ae. 37 t.

2833 (21270) I   ICHV   BN |2834 (21267) I   ICHV   BN

AT5   ROYAL SALOONS

Non-standard livery: All Royal vehicles are in Royal purple.

AT51. Royal Saloon (Royal family or household).

Dia AT501. Mark 1. This vehicle is a side corridor with a lounge, four bedrooms and a bathroom. Air conditioned. Converted 1977 from vs to ae and B5 bogies.

Lot No. 30130 Wolverton 1955. ETH 5X. 42 t.

2900 **0**   ICHX   ZN

AT51. Royal Saloon (Private secretary and Royal household).

Dia AT503. Mark 1. This vehicle is a side corridor with an attendant's compartment, three bedrooms two bathrooms and a lounge/office compartment. Air conditioned. Converted 1977 from vs to ae and B5 bogies.

Lot No. 30131 Wolverton 1957. ETH 5X. 42 t.

2901 **0**   ICHX   ZN

AT5G. The Queen's Saloon.

Dia. AT525. Mark 3. Converted from a mark 3 FO built 1972. Consists of a lounge, bedroom and bathroom for the Queen, and a combined bedroom and bathroom for the Queen's dresser. One entrance vestibule has double doors. Air conditioned. ae. BT10 bogies.

Lot No. 30886 Wolverton 1977. ETH 9X. 36 t.

2903 (11001) **0**   ICHX   ZN

AT5G. The Duke of Edinburgh's Saloon.

Dia. AT526. Mark 3. Converted from a mark 3 TSO built 1972. Consists of a combined lounge/dining room, a bedroom and a shower room for the Duke, a kitchen and a valet's bedroom and bathroom. Air conditioned. ae. BT10 bogies.

Lot No. 30887 Wolverton 1977. ETH 15X. 36 t.

2904 (12001) **0**   ICHX   ZN

AT5B. Staff Couchette/Power Brake.

Dia. AT527. Mark 2B. Converted from a Mk. 2B BFK built 1969. Consists of luggage accommodation, guard's compartment, 350 kW diesel generator and Staff sleeping accommodation. Pressure ventilated. ae. B5 bogies.

Lot No. 30888 Wolverton 1977. ETH 5X. 46 t.

2905 (14105) **0**   ICHX   ZN

AT5B. Staff Couchette.

Dia. AT528. Mark 2B. Converted from a Mk. 2B BFK built 1969. Pressure ventilated. ae. B5 bogies.

Lot No. 30889 Wolverton 1977. ETH 4X. 35.5 t.

2906 (14112) **0**   ICHX   ZN

AT5G. Royal Train Staff Sleeping Cars.

Dia. AT531. Mark 3A Details as for 10646 – 732 except that controlled emission toilets are not fitted. ETH11X.

Lot No. 31002 Derby/Wolverton 1985. 42.5 t 2915 is 44 t.

2914   **0**   ICHX   ZN |2915   **0**   ICHX   ZN

AT5G Royal Dining Car.

Lot No. 31059 Wolverton 1986. Converted from HST TRUK. Dia. AT537.

2916 (40512) **0**   ICHX   ZN

AT5G New Royal Vehicles. Full details not available.

Lot Nos. 31084 Wolverton 1989. Dia. AT539. 43 t. ETH13X.

2917 (40514) **0**   ICHX   ZN
Lot Nos. 31083 Wolverton 1989. Dia. AT538. 41.05 t. ETH10X.

2918 (40515) **0**   ICHX   ZN
Lot Nos. 31085 Wolverton 1989. Dia. AT540.

2919 (40518) **0**   ICHX   ZN

AT5B. Royal Train Staff/Generator Vehicle. Dia. AT536. Mark 2B. B5 bogies. 48 t. ETH2X.

Lot Nos. 31044 Wolverton 1986.

2920 (17109) **0**   ICHX   ZN

AT5B. Royal Staff Couchette. Dia. AT541. Mark 2B. B4 bogies. Dia. AT541. 41.5 t. ETH7X.

Lot No. 31086 Wolverton 1990.

2921 (17107) **0**   ICHX   ZN

AT5G. Royal Sleeping Car.

Lot No. 31035 Derby/Wolverton 1987. Dia. AT534.

2922 **0**   ICHX   ZN

AT5G. The Prince of Wales's Saloon.

Lot No. 31036 Derby/Wolverton 1987. Dia. AT535.

2923 **0**   ICHX   ZN

AD11 (FO)   OPEN FIRST

Dia. AD103. Mark 1. 42F 2L. ETH 3. d. Now fitted with table lights for use in first class charter trains. pa.

Lot No. 30576 BRCW 1959. B4 bogies. 33 t.

| 3097 a | I | ICHV | BN | 3100 x | I | ICHV | BN |
| 3098 a | I | ICHV | BN | | | | |

Later design with fluorescent lighting, aluminium window frames and Commonwealth bogies. pa.

3107 – 3127. Lot No. 30697 Swindon 1962 – 3. 36 t.
3131 – 3150. Lot No. 30717 Swindon 1963. 36 t.

Note: 3136/41/3/4/6/7/8 were renumbered 1060/3/5/6/8/9/70 when reclassified RUO, then 3605/8/9/2/6/4/10 when declassified, but have now regained their original numbers.

Note: 3131 – 3 are leased to Flying Scotsman Railways (see FSS 99190 – 2).

3107 x	I	ICHV	BN	3133 x	M	ICHL	BN
3111 x	I	ICHV	BN	3134 x	I	ICHV	BN
3114 x	I	ICHV	BN	3136 a	I	ICHV	BN
3115 x	I	ICHV	BN	3140 x	I	ICHV	BN
3118 x	I	ICHV	BN	3141 a	I	ICHV	BN
3119 x	I	ICHV	BN	3143 a	I	ICHV	BN
3120 x	I	ICHV	BN	3144 a	I	ICHV	BN
3121 a	I	ICHV	BN	3146 a	I	ICHV	BN
3123 a	I	ICHV	BN	3147 a	I	ICHV	BN
3124 a	I	ICHV	BN	3148 a	I	ICHV	BN
3127 a	I	ICHV	BN	3149 a	I	ICHV	BN
3131 x	M	ICHL	BN	3150 a	I	ICHV	BN
3132 x	M	ICHL	BN				

AD1D (FO)   OPEN FIRST

Dia. AD105. Mark 2D. Air conditioned. 3172 – 88 have Stones equipment. 3192/3202 have Temperature Ltd and require at least 800 V train heating supply. 42F 2L. B4 bogies. ae. pa. ETH 5.

Lot No. 30821 Derby 1971 – 2. 32.5 t.

3172	I	ICHX	CL	3186	I	ICHX	CL
3174	I	ICHX	CL	3187	I	ICHX	CL
3178	I	ICHX	CL	3188	I	ICHX	CL
3181	I	ICHV	BN	3192	I	ICHH	HQ
3182	I	ICHX	CL	3202	I	ICHH	HQ

AD1E (FO)   OPEN FIRST

Dia. AD106. Mark 2E. Air conditioned. Stones equipment. Require at least 800 V train heating supply. 42F 2L (41F 2L w). B4 bogies. ae. pa. ETH 5.

* Seats removed to accommodate catering module. 40F 1L.

§ Fitted with power supply for Mk. 1 RBR.

Lot No. 30843 Derby 1972 – 3. 32.5 t.

3221 w	I	IANR	NC	3247	I	ICHV	BN
3223	I	ICHH	HQ	3248	I	ICHX	CL
3224	I	ICCX	DY	3249 *	I	IWCX	WB
3225	I	ICHH	HQ	3250 w	I	ICHV	BN
3226	I	ICHH	HQ	3251 *	I	ICHH	CL
3227	I	ICHV	BN	3252 w	I	ICCX	DY
3228 §	I	IANR	NC	3256 w	I	ICCX	DY
3229	I	ICCX	DY	3257 w	I	ICHX	CL
3230	I	ICHX	CL	3258 n	I	ICHH	HQ
3231	I	ICHV	BN	3259 *	I	IWCX	WB
3232 w	I	ICCL	DY	3261 w	I	ICCX	MA
3233		ICHH	CL	3262	I	ICHX	BN
3234 w	I	ICHV	BN	3263	I	ICHV	BN
3235 §	I	IANX	NC	3265	I	ICHV	BN
3237	I	ICHV	BN	3266 §	I	ICHX	BN
3239	I	ICHX	CL	3267	I	ICHX	CL
3240	I	ICHX	CL	3268	I	ICHH	HQ
3241	I	ICCX	PC	3269	I	ICCX	DY
3242 w§I		ICCX	DY	3270	I	ICHV	BN
3244 w	I	IANR	NC	3272	I	ICHV	BN
3245 w	I	ICHH	HQ	3273	I	ICHX	CL
3246 w	I	ICHX	CL	3275	I	ICHV	BN

AD1F (FO)   OPEN FIRST

Dia. AD107. Mark 2F. Air conditioned. 3277 – 3318/58 – 81 have Stones equipment, others have Temperature Ltd. 42F 2L. All now refurbished with power-operated vestibule doors, new panels and new seat trim. B4 bogies. ae. pa. ETH 5X.

3277 – 3318. Lot No. 30845 Derby 1973. 33 t.
3325 – 3428. Lot No. 30859 Derby 1973 – 4. 33 t.
3429 – 3438. Lot No. 30873 Derby 1974 – 5. 33 t.

§ Fitted with power supply for Mk. 1 RBR.

3277	I	IANR	NC	3313	I	IWCR	OY
3278	I	IWCR	OY	3314	I	IWCR	MA
3279 §	I	IANR	NC	3318	I	IANR	NC
3285	I	IWCR	MA	3325	I	IWCR	OY
3290	I	IANR	NC	3326	I	IWCR	OY
3292	I	IANR	NC	3330	I	IWCR	OY
3293	I	IWCR	MA	3331	I	IANR	NC
3295	I	IWCR	OY	3333	I	IWCR	OY
3299	I	IWCR	OY	3334	I	IANR	NC
3300	I	IWCR	OY	3336 §	I	IANR	NC
3303	I	IANR	NC	3337	I	IWCR	OY
3304	I	IWCR	OY	3338 §	I	IANR	NC
3309	I	IANR	NC	3340	I	IWCR	OY
3312	I	IWCR	MA	3344	I	IWCR	OY

3345	I	IWCR	OY	3387	I	IWCR	OY
3348	I	IWCR	OY	3388	I	IANR	NC
3350	I	IWCR	OY	3389	I	IWCR	MA
3351	I	IANR	NC	3390	I	IWCR	OY
3352	I	IWCR	OY	3392	I	IWCR	OY
3353	I	IWCR	OY	3395	I	IWCR	OY
3354	I	IWCR	OY	3397	I	IWCR	OY
3356	I	IWCR	MA	3399 §	I	IANR	NC
3358	I	IANR	NC	3400	I	IANR	NC
3359	I	IWCR	MA	3402	I	IWCR	OY
3360	I	IWCR	OY	3403	I	IWCR	OY
3362	I	IWCR	OY	3408	I	IWCR	OY
3363	I	IWCR	OY	3411	I	IWCR	OY
3364	I	IWCR	MA	3414	I	IANR	NC
3366	I	IWCR	OY	3416	I	IANR	NC
3368	I	IANR	NC	3417	I	IANR	NC
3369	I	IWCR	OY	3424	I	IWCR	OY
3373	I	IWCR	OY	3425	I	IWCR	OY
3374	I	IWCR	OY	3426	I	IWCR	OY
3375	I	IANR	NC	3428	I	IWCR	OY
3379 §	I	IANR	NC	3429	I	IWCR	OY
3381	I	IWCR	OY	3431	I	IWCR	OY
3384	I	IWCR	OY	3433	I	IWCR	OY
3385	I	IWCR	OY	3434	I	IWCR	OY
3386	I	IWCR	OY	3438	I	IWCR	OY

AG1E (FO)   OPEN FIRST (PANTRY)

Dia. AG101. Mark 2E. Air conditioned. Converted from FO. Fitted with pantry, microwave oven and payphone for use on sleeping car services. 36F 1L. B4 bogies. ae. pa. ETH 5X.

Lot No. 30843 Derby 1972 – 3. 32.5 t.

3520 (3253)	I	IWRX	LA	3523 (3238)	I	IWCX	WB
3521 (3271)	I	IWRX	LA	3524 (3254)	I	IWCX	WB
3522 (3236)	I	IWRX	LA	3525 (3255)	I	IWCX	WB

AC21 (TSO)   OPEN STANDARD

Dia. AC201. Mark 1. These vehicles have 2 + 2 seating and are classified TSO ('Tourist second open' – a former LNER designation). They are a development of Dia. AC204 (all now withdrawn) with fluorescent lighting and modified design of seat headrest. 64S 2L. ETH 4.

Note: 4860 and 5032/5 are leased to Flying Scotsman Railways (see FSS 99193 – 5).

Lot No. 30646 Wolverton 1961. Built with Commonwealth bogies, but BR1 bogies substituted by the SR on 4902/5/9/10/15/16. All now re-rebogied. 34 t B4, 36 t C.

4842 x	I	ICHD	CL	4854 v pa	**NR** RCXX		LL
4849 v pa	**NR** RCXX		LL	4858 x	I	ICHH	CL

4860 x	**M** ICHL	BN		4875 v pa	**NR** RCXX	LL
4866 v pa	**N** RCXX	LL		4876 v pa	**NR** RCXX	LL
4869 x	**I** ICHD	CL		4880 v pa	**NR** RCXX	LL
4873 v pa	**NR** RCXX	LL				

Lot No. 30690 Wolverton 1961 – 2. Commonwealth bogies and aluminium window frames. d. 37 t.

f – Facelifted with new laminate, new diffusers etc.

4902 x B4	**I** ICHH	CL		4984 v	**E** ICHS	CL
4905 vCpa	**NI** ICHS	CL		4986 a pa	**I** ICHD	CL
4909 x B4	**I** ICHD	CL		4991 a pa	**I** ICHD	CL
4910 vCpa	**NI** ICHS	CL		4993 a pa	**I** ICHD	CL
4915 x B4	**I** ICHD	CL		4994 v pa	**NI** ICHS	CL
4916 x B4	**I** ICHD	CL		4996 afpa	**I** RFXZ	HQ
4917 vCpa	**RR** RCXX	LL		4998 a pa	**I** ICHD	CL
4925 a pa	**I** ICHD	CL		4999 a pa	**I** ICHD	CL
4936 v pa	**NI** ICHS	CL		5002 a pa	**I** ICHD	CL
4938 a pa	**I** ICHD	CL		5005 a pa	**I** ICHD	CL
4939 a pa	**I** ICHD	CL		5007 a pa	**I** ICHD	CL
4940 v pa	**NI** ICHS	CL		5010 a pa	**I** ICHD	CL
4946 afpa	**I** RFXZ	HQ		5023 a pa	**I** ICHD	CL
4949 a pa	**I** ICHD	CL		5027 a pa	**I** ICHD	CL
4951 v pa	**NI** ICHS	CL		5032 x	**M** ICHL	BN
4956 a pa	**I** ICHD	CL		5035 x	**M** ICHL	BN
4959 a pa	**I** ICHD	CL		5037 a pa	**I** ICHD	CL
4960 v	**E** ICHS	CL		5040 x	**I** ICHH	CL
4973 v pa	**NI** ICHS	CL		5041 a pa	**I** ICHD	CL
4977 a pa	**I** ICHD	CL		5042 x	**I** ICHD	CL
4979 afpa	**I** RFXZ	HQ				

AC2Z (TSO)   OPEN STANDARD

Dia. AC205. Mark 2. Pressure ventilated. 64S 2L. B4 bogies. vd. ETH 4.

Lot No. 30751 Derby 1965 – 7. 32 t.

5104	**RR** RAIS	IS		5174 pa	**RR** RAIS	IS
5132 pa	**E** RAIS	IS		5177	**RR** RAIS	IS
5133	**RR** RAIS	IS		5179	**RR** RAIS	IS
5135	**RR** RAIS	IS		5180	**RR** RAIS	IS
5138	**RR** RAIS	IS		5183	**RR** RAIS	IS
5139 pa	**E** RAIS	IS		5184	**RR** RAIS	IS
5148 pa	**RR** RAIS	IS		5186	**RR** RAIS	IS
5154 pa	**E** RAIS	IS		5191 pa	**E** RAIS	IS
5156 pa	**RR** RAIS	IS		5193 pa	**E** RAIS	IS
5157	**RR** RAIS	IS		5194	**RR** RAIS	IS
5158	**RR** RAIS	IS		5198	**RR** RAIS	IS
5159	**RR** RAIS	IS		5207 pa	**RR** RAXX	IS
5161	**RR** RAXX	IS		5209	**RR** RAIS	IS
5166 pa	**E** RAIS	IS		5210	**RR** RAXX	IS
5167	**RR** RAIS	IS		5212 pa	**E** RAIS	IS
5173	**RR** RAIS	IS		5213	**RR** RAXX	IS

5221 pa	**RR** RAIS	IS		5226	**RR** RAXX	IS	
5225	**RR** RAXX	IS					

Named vehicles:

5132 CLAN MUNRO	5191 CLAN DONALD
5139 CLAN ROSS	5193 CLAN MACLEOD
5154 CLAN FRASER	5212 CAPERKAILZIE
5166 CLAN MACKENZIE	

AD2Z (SO)   OPEN STANDARD

Dia. AD203. Mark 2. Pressure ventilated. 48S 2L. B4 bogies. Originally used as restaurant cars. d. pa. ETH 4.

* Converted to SO(T) seating 40S 1L.

Lot No. 30752 Derby 1966.

5230 v* **E**	RAIS	IS	5234 v* **E**	RAIS	IS
5233 v	**RR** RAIS	IS			

Named vehicles:

5230 CORRIEMOILLIE	5234 CLAN MACKINTOSH

AC2A (TSO)   OPEN STANDARD

Dia. AC206. Mark 2A. Pressure ventilated. 64S 2L. B4 bogies. ad. ETH 4.

5265 – 5341. Lot No. 30776 Derby 1967 – 8. 32 t.
5350 – 5433. Lot No. 30787 Derby 1968. 32 t.

5265 pa	**NR** RCLL	LL	5331 pa	**NR** RCLL	LL
5266	**RR** RCLL	LL	5335	**NR** RCLL	LL
5267	**RR** RCLL	LL	5337	RBHT	HT
5271 pa	**RR** RCLL	LL	5341	**RR** RCLL	LL
5272 pa	**RR** RCLL	LL	5345 pa	**RR** RCLL	LL
5275	**NR** RCLL	LL	5350 pa	**NR** RCLL	LL
5276 pa	**NR** RCLL	LL	5353 pa	**RR** RCLL	LL
5277 pa	RBHT	HT	5354	**RR** RCLL	LL
5278 pa	**NR** RCLL	LL	5362	**RR** RCXX	LL
5279 pa	RBHT	HT	5364	**RR** RCLL	LL
5282 pa	**RR** RCLL	LL	5365	**RR** RCLL	LL
5290 pa	**NR** RCLL	LL	5366	**RR** RCLL	LL
5291	**RR** RCLL	LL	5373	**NR** RCLL	LL
5292 pa	**RR** RCLL	LL	5376	**NR** RCLL	LL
5293	**NR** RCLL	LL	5378	**NR** RCLL	LL
5300 pa	RBHT	HT	5379	**RR** RCLL	LL
5304	**RR** RCLL	LL	5381 pa	**NR** RCLL	LL
5307 pa	**RR** RCLL	LL	5384 pa	**N** RBHT	HT
5309	**NR** RCLL	LL	5386	**RR** RCLL	LL
5314	RBHT	HT	5389 pa	**NR** RCLL	LL
5316	**RR** RCLL	LL	5392 pa	RBHT	HT
5322	**RR** RCLL	LL	5393 pa	**RR** RCLL	LL
5323	**RR** RCLL	LL	5396	**RR** RCLL	LL

5401 pa **RR** RCLL	LL	5420 pa **NR** RCLL	LL
5412 **NR** RCLL	LL	5433 pa **NR** RCLL	LL
5419 pa **NR** RCLL	LL		

AC2D (TSO)   OPEN STANDARD

Dia. AC209. Mark 2D. Air conditioned. Stones (5653 has Temperature Ltd.) equipment. 62S 2L. B4 bogies. ae. pa. ETH 5.

Lot No. 30822 Derby 1971. 33 t.

5616	I	ICHX	BN	5676	I	ICCR	PC
5617	I	ICHH	HQ	5679	I	ICCR	MA
5618	I	ICCR	PC	5682	I	ICCR	PC
5620	I	ICCR	PC	5685	I	ICCR	PC
5621	I	IXXB	HQ	5686	I	ICCR	PC
5623	I	ICCR	PC	5687	I	IXXT	HQ
5624	I	ICCR	PC	5689	I	ICHX	BN
5625	I	IANX	NC	5690	I	ICCR	PC
5626	I	ICCR	PC	5692	I	ICCR	PC
5628	I	ICCR	MA	5693	I	IANR	NC
5629	I	ICCR	PC	5694	I	IXXT	HQ
5630	I	ICHX	BN	5695	I	IANR	NC
5631	I	ICCR	PC	5699	I	IXXH	HQ
5632	I	ICCR	PC	5700	I	ICCR	PC
5633	I	ICHH	HQ	5701	I	IXXT	HQ
5634	I	ICCR	MA	5703	I	ICCR	MA
5636	I	ICCR	PC	5705	I	IANX	NC
5637	I	ICHH	HQ	5706	I	IXXZ	HQ
5638	I	ICCR	PC	5710	I	ICCR	MA
5639	I	IXXB	HQ	5711	I	ICCR	PC
5640	I	IXXH	HQ	5715	I	ICCR	PC
5646	I	ICCR	PC	5716	I	IXXT	HQ
5647	I	ICHX	BN	5717	I	IXXT	HQ
5648	I	ICHH	HQ	5718	I	IXXT	HQ
5650	I	ICCR	PC	5719	I	IANX	NC
5651	I	ICCR	PC	5722	I	ICHX	CL
5652	I	ICCR	PC	5723	I	ICCR	PC
5653	I	ICCR	PC	5724	I	ICCR	PC
5654	I	ICCR	MA	5726	I	ICCR	PC
5657	I	ICCR	PC	5728	I	IANR	NC
5658	I	ICCR	PC	5729	I	ICHH	HQ
5659	I	IXXB	HQ	5730	I	ICCR	PC
5660	I	IXXT	HQ	5731	I	IXXT	HQ
5661	I	IXXT	HQ	5732	I	ICHX	CL
5662	I	ICCR	PC	5734	I	ICCR	PC
5663	I	IXXT	HQ	5735	I	ICCR	MA
5665	I	ICCR	MA	5737	I	ICCR	MA
5669	I	ICCR	MA	5738	I	IXXT	HQ
5671	I	ICCR	MA	5739	I	ICHX	CL
5673	I	ICCR	PC	5740	I	ICCR	PC
5674	I	IXXT	HQ	5742	I	IXXZ	HQ
5675	I	ICHH	HQ	5743	I	ICCR	MA

Mark 3 Royal train saloon No. 2903 is pictured at Andover on 25th June 1993.

Brian Denton

▲ **Mark 1 Stock.** Intercity liveried minature buffet car (RMB) No. 1842 in the formation of the 09.05 Paddington – Manchester Piccadilly on 5th September 1993.   *David Brown*

▼ LNER tourist green and cream liveried open standard (TSO) No. 4960 is pictured at Gretna on 25th March 1993 in an empty coaching stock working.
                                                                    Kevin Conkey

Stored blue & grey corridor first (FK) No. 13306 at Carlisle Upperby on 27th February 1993. *Kevin Conkey*

Corridor brake standard (BSK) No. 35479 at Carlisle Upperby on 15th October 1993. This vehicle carries a form of modified Network SouthEast livery, the blue being replaced with dark grey for use on Intercity services.

Kevin Conkey

Mark 2 Stock. Regional Railways open standard (SO) No. 5233 at Kyle of Lochalsh on 26th September 1992.

John Augustson

▲ **Mark 2A Stock.** Open standard (TSO) No. 5384 in Network SouthEast livery at Hartlepool on 24th July 1993. *John Augustson*

▼ **Mark 2D Stock.** Corridor brake first (BFK) No. 17169 at Carlisle on 20th March 1993. *Kevin Conkey*

▲ **Mark 2E Stock.** Open first (FO) No. 3240 at Carlisle Upperby on 23rd September 1993.
Kevin Conkey

▼ **Mark 2F Stock.** Driving open brake standard No. 9709 leading the 09.30 Liverpool Street – Norwich at Colchester on 23rd October 1992.
John Augustson

▲ **Mark 3 Stock.** HST Trailer buffet standard (TRSB) No. 40417 at Durham on 7th May 1993. *John Augustson*

▼ **Mark 3A Stock.** Sleeping car No. 10655 in the formation of a charter train at Hartlepool on 10th April 1993. *John Augustson*

AC2E (TSO) OPEN STANDARD

Dia. AC210. Mark 2E. Air conditioned. Stones equipment. 64S 2L (62S 2L w).
B4 bogies. Require at least 800 V train heat supply. ae. pa. ETH 5.

5744 – 5803. Lot No. 30837 Derby 1972. 33.5 t.
5810 – 5907. Lot No. 30844 Derby 1972 – 3. 33.5 t.

5744	I	ICCX	MA	5801	I ICCR	PC
5745	I	IXXT	HQ	5803	I ICHX	BN
5746	I	ICCR	PC	5810	I IXXT	HQ
5747	I	ICHX	BN	5811	I ICHX	CL
5748 w	I	ICCR	PC	5812 w	I ICCR	PC
5750	I	IXXT	HQ	5814	I ICCR	PC
5751 w	I	ICCR	PC	5815	I IXXH	HQ
5752 w	I	ICCX	PC	5816	I ICCR	PC
5754 w	I	ICCR	PC	5818	I ICHX	BN
5755	I	IXXH	HQ	5820	I ICHX	BN
5759	I	ICHX	CL	5821	I ICCR	PC
5760	I	ICCR	PC	5822	I ICCX	MA
5761	I	ICHX	CL	5823	I IXXZ	HQ
5762	I	ICHX	CL	5824 w	I ICCR	PC
5763	I	IXXT	HQ	5826	I ICCR	PC
5764	I	ICCX	MA	5827 w	I ICCX	MA
5766	I	ICCX	MA	5828 w	I ICCX	MA
5768	I	IXXT	HQ	5829	I ICHX	CL
5769	I	ICCR	PC	5831	I ICHX	CL
5770	I	ICHX	BN	5832	I IXXZ	HQ
5772 w	I	ICCR	PC	5833	I ICCX	MA
5773	I	IXXT	HQ	5834	I IXXT	HQ
5775	I	IXXT	HQ	5835	I ICHX	CL
5776	I	ICCR	PC	5836	I ICHX	BN
5777	I	ICHX	CL	5837	I ICHX	CL
5778 w	I	ICCR	PC	5838	I IXXT	HQ
5779	I	ICCX	MA	5840	I ICCX	MA
5780 w	I	ICCR	PC	5842 w	I ICHX	CL
5781 w	I	ICCX	MA	5843 w	I ICCR	PC
5784	I	ICCR	PC	5844	I IXXB	HQ
5785	I	IXXT	HQ	5845 w	I ICCX	MA
5786	I	ICHH	HQ	5847 w	I ICCR	PC
5787	I	IXXT	HQ	5849	I IXXT	HQ
5788 w	I	ICCX	MA	5851	I ICCX	MA
5789	I	ICCR	PC	5852	I ICHX	CL
5791 w	I	ICCR	PC	5853	I ICCX	MA
5792	I	ICCR	PC	5854	I ICCX	MA
5793	I	ICCR	PC	5859	I ICCX	MA
5794 w	I	ICCR	PC	5860 w	I ICHX	CL
5795	I	IXXT	HQ	5861	I ICHX	BN
5796 w	I	ICCR	MA	5863	I ICHX	BN
5797	I	ICCR	PC	5866	I ICCX	MA
5799	I	ICCX	PC	5868	I ICCR	PC
5800	I	ICHX	CL	5869	I ICCR	PC

5870	I	IXXT	HQ	5889	I	ICCR	PC
5871	I	ICCX	MA	5890	I	ICCX	MA
5872 n	I	IXXZ	HQ	5891	I	ICHX	BN
5873	I	ICHX	CL	5892	I	ICCR	PC
5874 w	I	ICCX	PC	5893	I	ICCR	PC
5875	I	ICCX	MA	5897	I	ICCR	MA
5876	I	ICCR	PC	5899	I	ICCR	PC
5878	I	ICHX	BN	5900	I	ICCR	PC
5879	I	ICHH	HQ	5901	I	IXXT	HQ
5881	I	IXXT	HQ	5902	I	ICCX	MA
5883	I	ICHX	CL	5903	I	ICCX	MA
5884	I	IXXB	HQ	5904	I	IXXZ	HQ
5886	I	IXXT	HQ	5905	I	ICCR	PC
5887 w	I	ICCR	PC	5906	I	ICCR	MA
5888 w	I	ICCR	MA	5907	I	ICHX	CL

AC2F (TSO)   OPEN STANDARD

Dia. AC211. Mark 2F. Air conditioned. Temperature Ltd. equipment. 64S 2L. Inter-City 70 seats. All now refurbished with power-operated vestibule doors, new panels and new seat trim. B4 bogies. ae. pa. ETH 5X.

* – Early Mark 2 style seats.

5908 – 5958. Lot No. 30846 Derby 1973. 33 t.
5959 – 6170. Lot No. 30860 Derby 1973 – 4. 33 t.
6171 – 6184. Lot No. 30874 Derby 1974 – 5. 33 t.

w Wheelchair space. 62S 2L.

5908	I	IWCR	OY	5934	I	IWCR	OY
5910 w	I	IWCR	OY	5935	I	IANR	NC
5911	I	IWCR	MA	5936	I	IANR	NC
5912	I	ICCR	DY	5937	I	IWCR	OY
5913	I	ICCR	MA	5939	I	IWCR	OY
5914	I	IWCR	OY	5940 w	I	IWCR	MA
5915	I	IWCR	OY	5941	I	IWCR	OY
5916 w	I	ICCR	DY	5943 w	I	IWCR	OY
5917	I	ICCR	MA	5944 w	I	IWCR	OY
5918 w	I	ICCR	DY	5945 w	I	IWCR	MA
5919	I	IWCR	OY	5946	I	IWCR	OY
5920	I	IWCR	OY	5947	I	ICCR	MA
5921	I	IANR	NC	5948 w	I	IWCR	OY
5922	I	IANR	NC	5949 w	I	IWCR	MA
5924	I	IANR	NC	5950	I	IANR	NC
5925 w	I	ICCR	MA	5951	I	ICCR	MA
5926	I	IANR	NC	5952	I	IWCR	MA
5927	I	IANR	NC	5953	I	IWCR	OY
5928	I	IANR	NC	5954	I	IANR	NC
5929	I	IANR	NC	5955	I	IWCR	OY
5930 w	I	ICCR	MA	5956	I	IANR	NC
5931 w	I	IWCR	OY	5957	I	IWCR	OY
5932	I	IWCR	OY	5958	I	IWCR	OY
5933	I	IWCR	OY	5959 n	I	IANR	NC

No.		Code	State		No.		Code	State
5960	I	IWCR	MA		6026 *	I	ICCR	MA
5961	I	ICCR	MA		6027 w	I	IWCR	OY
5962	I	ICCR	DY		6028	I	IWCR	OY
5963	I	IWCR	OY		6029	I	IWCR	OY
5964	I	IANR	NC		6030 w	I	ICCR	MA
5965 w	I	ICCR	DY		6031	I	IWCR	OY
5966	I	IANR	NC		6034	I	IWCR	OY
5967 w	I	ICCR	MA		6035 w	I	ICCR	DY
5968	I	IANR	NC		6036 *	I	IANR	NC
5969 w	I	IWCR	OY		6037	I	IANR	NC
5971	I	ICCR	MA		6038	I	ICCR	MA
5973	I	IANR	NC		6041	I	ICCR	MA
5975 * -	I	ICCR	DY		6042	I	IANR	NC
5976 w	I	ICCR	MA		6043	I	IWCR	MA
5977	I	IWCR	MA		6045 w	I	IWCR	OY
5978 *	I	IWCR	MA		6046	I	IWCR	OY
5980	I	IWCR	OY		6047 *n	I	IWCR	OY
5981	I	ICCR	MA		6049	I	IWCR	MA
5983	I	ICCR	MA		6050	I	ICCR	MA
5984 *	I	IWCR	OY		6051 *	I	IWCR	OY
5985	I	IWCR	OY		6052 w	I	ICCR	MA
5986	I	IWCR	OY		6053 *	I	IANR	NC
5987 *	I	IWCR	OY		6054	I	IWCR	OY
5988 w	I	IWCR	OY		6055	I	IWCR	OY
5989 w	I	ICCR	MA		6056	I	IWCR	OY
5991	I	ICCR	MA		6057	I	IWCR	OY
5993 *w	I	IANR	NC		6059	I	ICCR	MA
5994 *	I	ICCR	MA		6060 *	I	IWCR	OY
5995	I	ICCR	DY		6061 *	I	ICCR	MA
5996	I	IWCR	OY		6062 *	I	IWCR	OY
5997	I	IWCR	OY		6063 w	I	IWCR	OY
5998	I	IANR	NC		6064	I	IWCR	OY
5999	I	ICCR	MA		6065	I	IWCR	OY
6000	I	IWCR	MA		6066	I	ICCR	MA
6001 w	I	IWCR	OY		6067	I	IWCR	OY
6002	I	IWCR	OY		6073	I	ICCR	MA
6005 *	I	ICCR	MA		6100 *	I	IWCR	MA
6006	I	IWCR	MA		6101	I	IWCR	OY
6008	I	ICCR	MA		6102	I	IWCR	MA
6009	I	IWCR	OY		6103	I	IANR	NC
6010 n	I	ICCR	MA		6104	I	IWCR	OY
6011	I	ICCR	MA		6105	I	ICCR	DY
6012 *	I	IWCR	OY		6106	I	IWCR	MA
6013 *	I	ICCR	DY		6107	I	IWCR	MA
6014	I	ICCR	MA		6110 w	I	IANR	NC
6015 w	I	ICCR	MA		6111	I	IWCR	OY
6016	I	IWCR	OY		6112	I	ICCR	MA
6018 *	I	IWCR	OY		6113	I	IWCR	OY
6021	I	IWCR	OY		6115	I	ICCR	MA
6022 w	I	ICCR	DY		6116	I	IWCR	OY
6024	I	ICCR	MA		6117 w	I	ICCR	MA
6025 *w	I	ICCR	MA		6119 w	I	ICCR	DY

6120	I	ICCR	MA		6158 *	I	IWCR	OY
6121	I	IWCR	MA		6159 *n	I	ICCR	MA
6122	I	IWCR	OY		6160 *	I	IANR	NC
6123	I	IANR	NC		6161 *	I	IWCR	OY
6124	I	ICCR	MA		6162	I	ICCR	MA
6134	I	IWCR	OY		6163	I	IWCR	MA
6135	I	ICCR	DY		6164	I	IWCR	OY
6136	I	IWCR	OY		6165	I	IWCR	OY
6137	I	ICCR	MA		6166	I	IANR	NC
6138	I	IWCR	OY		6167	I	IANR	NC
6139 *n	I	IANR	NC		6168	I	ICCR	DY
6141 w	I	IWCR	OY		6170	I	IWCR	OY
6142 *	I	IWCR	OY		6171	I	IWCR	OY
6144 *	I	IWCR	OY		6172	I	ICCR	MA
6145 *	I	ICCR	MA		6173	I	IWCR	OY
6146 *	I	IWCR	OY		6174	I	IANR	NC
6147 *	I	IWCR	MA		6175	I	IWCR	OY
6148 *	I	ICCR	DY		6176 w	I	ICCR	MA
6149 *w	I	IWCR	OY		6177	I	ICCR	MA
6150 *	I	ICCR	DY		6178 w	I	IWCR	OY
6151 *	I	IWCR	OY		6179	I	IWCR	OY
6152 *	I	IANR	NC		6180 w	I	IWCR	OY
6153 *	I	IWCR	MA		6181 wn	I	IWCR	OY
6154 *	I	ICCR	MA		6182	I	ICCR	DY
6155 *	I	IANR	NC		6183	I	ICCR	MA
6157 *	I	ICCR	MA		6184 *	I	ICCR	MA

AC2D (TSO)   OPEN STANDARD

Dia. AC217. Mark 2D. Air conditioned. Stones. 58S 2L. (58S 1L*). B4 bogies.
ae. pa. ETH 5X. Rebuilt from FO with new style 2 + 2 seats.

Lot No. 30821 Derby 1971 – 2. 33.5 t.

6200 (3198)	I IWRX	LA		6218 (3209)	I IWCX	WB
6201 (3210) *I	IWCX	WB		6219 (3213)	I IWCX	WB
6202 (3191) *I	IWCX	WB		6220 (3175)	I IWCX	WB
6203 (3180)	I IWCX	WB		6221 (3173)	I IWCX	WB
6204 (3216)	I ICHX	BN		6222 (3171)	I IWCX	WB
6205 (3193)	I ICCX	DY		6223 (3194)	I ICHX	BN
6206 (3183)	I IWRX	LA		6224 (3195) *I	IWCX	WB
6207 (3204)	I ICCX	DY		6225 (3200)	I ICHX	BN
6208 (3205)	I IWCX	WB		6226 (3203)	I IWRX	LA
6209 (3177)	I ICHX	BN		6227 (3197)	I ICCX	DY
6210 (3196) *I	IWCX	WB		6228 (3201) *I	IWCX	WB
6211 (3215)	I IWCX	WB		6229 (3212)	I IWCX	WB
6212 (3176)	I IWCX	WB		6230 (3185)	I ICCX	DY
6213 (3208)	I IWRX	LA		6231 (3189)	I ICHX	BN
6214 (3211)	I IWCX	WB		6232 (3199) *I	IWCX	WB
6215 (3170)	I ICCX	DY		6233 (3206)	I ICCX	DY
6216 (3179)	I IWCX	WB		6234 (3207)	I IWCX	WB
6217 (3184)	I IWCX	WB		6235 (3190)	I ICHX	BN

AD4Z (OC) — OBSERVATION CAR

Dia. AD401. Converted 1987 from DMU DTCL. DMU bogies. 42U 1L. v. pa.

Lot No. 30468 Metro-Cammell 1958. 25.5 t.

6300 (54356) E RAIS IS HEBRIDEAN

GX51 — GENERATOR VAN

Dia. GX501. Renumbered 1989 from BR departmental series. Three-phase supply generator van for use with HST trailers. Often used at times of low availability of HST power cars. Rebuilt from NDA 81448. B5 bogies.

Lot No. 30400 Pressed Steel 1958.   t.

6310 (ADB 975325) I IXXB PM

AX51 — GENERATOR VAN

Dia. AX501. Converted to generator vans for use with pairs of Class 37s in Scotland. B5 bogies.

6311. Lot No. 30162 Pressed Steel 1958.   t.
6312. Lot No. 30224 Cravens 1956.   t.
6313. Lot No. 30484 Pressed Steel 1958.   t.

6311 (80903, 92011, 92911) I IWCX IS
6312 (81023, 84023, 92025, 92925) I IWCX IS
6313 (81553, 84553, 92167) I IWCX IS

AZ5Z — SPECIAL SALOON

Dia. AZ501. Renumbered 1989 from LMR departmental series. Formerly the LMR General Manager's saloon. Rebuilt from LMS period 1 BFK M 5033 M to dia. 1654 and mounted on the underframe of BR suburban BS M 43232. B5 bogies. This vehicle has a maximum speed of 100 mph, but is restricted to 60 mph when carrying passengers with screw coupling operative.

LMS Lot No. 326 Derby 1927.   t.

6320 (TDM 395707) I ICHV BN

GS5(HSBV) — HST BARRIER VEHICLE

Various diagrams. Renumbered from departmental stock, or converted from various types. ae. B4 bogies (Commonwealth bogies *).

6330/9. Lot No. 30786 Derby 1968. 32 t.
6332/43. Lot No. 30795 Derby 1969/70. 32 t.
6334. Lot No. 30400 Pressed Steel 1957 – 8. 31.5 t.
6335. Lot No. 30775 Derby 1967 – 8. 32 t.
6336/8. Lot No. 30715 Gloucester 1962. 31 t.
6340. Lot No. 30669 Swindon 1962. 36 t.
6341/2. Lot No. 30632 Ashford/Swindon 1961. 38 t.

6343. Lot No. 30091 Doncaster 1954. 33 t.
6344. Lot No. 30163 Pressed Steel 1957. 31.5 t.
6345. Lot No. 30796 Derby 1970. 32.5 t.
6346. Lot No. 30777 Derby 1967. 31.5 t.
6347. Lot No. 30787 Derby 1968. 31.5 t.

6330 (14084, ADB 975629)	GS503		IWRG	LA
6332 (5594)	GS508	I	ICCG	EC
6334 (81478, 92128)	GS507	I	IMLG	NL
6335 (14065, ADB 975655)	GS503	I	ICCG	LA
6336 (81591, 92185)	GS507	I	IWRG	LA
6338 (81551, 92180)	GS507	I	IWRG	LA
6339 (14078, ADB 975666)	GS503	I	ICCG	NL
6340 (21251, ADB 975678)	GS504*	I	IWRG	LA
6341 (1967, ADB 975980)	GS505*	I	ILAG	OO
6342 (1983, ADB 975981)	GS505*	I	ILAG	OO
6343 (5522)	GS508	I	IMLG	NL
6344 (81263, 92080)	GS507	I	IECG	HT
6345 (14137, 17137)	GS510	I	ICCG	BN
6346 (9422)	GS511	I	ICCG	EC
6347 (5395)	GS509	I	IWRG	PM
6348 (92963)	GS507	I	IWRG	LA

GF5 (HSBV)   MARK 4 BARRIER VEHICLE

Various diagrams. Renumbered from departmental stock, or converted from FK or BSO. ae. B4 bogies.

6350. Lot No. 30472 BRCW 1959. 33 t.
6351. Lot No. 30091 Doncaster 1954. 33 t.
6352/3. Lot No. 30774 Derby 1968. 33 t.
6354 – 6. Lot No. 30820 Derby 1970. 32 t.
6357. Lot No. 30798 Derby 1970. 32 t.
6358 – 9. Lot No. 30788 Derby 1968. 31.5 t.
6390. Lot No. 30136 Metro-Cammell 1955. 31.5 t.

6350 (3088, ADB 977434)	AV501	I	IXXZ	HQ
6351 (3050, ADB 977435)	AV501		IECG	EC
6352 (13465, 19465)	AV502		IECG	BN
6353 (13478, 19478)	AV503		IECG	EC
6354 (9459)	AV504	I	IECG	BN
6355 (9477)	AV504		IECG	BN
6356 (9455)	AV504		IECG	BN
6357 (9443)	AV504		IECG	BN
6358 (9432)	AV505		IECG	BN
6359 (9429)	AV505		IECG	BN
6390 (92900)	AV506		IECG	BN

GF5 (BV)   DMU/EMU* BARRIER VEHICLE

Various diagrams. Converted from BFK, BSO or BG. ae. B4 bogies.

6360. Lot No. 30777 Derby 1967. 31.5 t.
6361 – 2. Lot No. 30820 Derby 1970. 32 t.

6363. Lot No. 30796 Derby 1970. 32 t.
6364. Lot No. 30039 Derby 1954. 32 t.
6365. Lot No. 30323 Pressed Steel 1957. 32 t.

6360 (9420)	**RR** RCXX	LL	6363 (17117)	**RR** RCXX	LL
6361 (9460)	**RR** RCXX	LL	6364 (80565) *	**RR** RCLG	LG
6362 (9467)	**RR** RCXX	LL	6365 (84296) *	**RR** RCLG	LG

AY5 (BV)　　　　EUROSTAR BARRIER VEHICLE

Dia. AY501. Converted from GUVs. Bodies removed to allow for nose of Eurostar set. a. B4 bogies.

6380 – 6382. Lot No. 30417 Pressed Steel 1958 – 9. 30 t.
6383. Lot No. 30565 Pressed Steel 1959. 30 t.

6380 (93386)	GPSM	HQ	6382 (93295)	GPSM	HQ
6381 (93187)	GPSM	HQ	6383 (93664)	GPSM	HQ

AG2D (TSOT)　　　OPEN STANDARD (TROLLEY)

Dia. AG202. Mark 2D. Converted from TSO by removal of one seating bay and replacing this by a counter with a space for a trolley. Adjacent toilet removed and converted to steward's washing area/store. Air conditioned. Stones equipment. 54S 1L. B4 bogies. ae. pa. ETH 5.

Lot No. 30822 Derby 1971. 33 t.

6605 (5741)	I	IXXT	HQ	6614 (5725)	I	IXXT	HQ
6608 (5696)	I	IXXT	HQ	6619 (5655)	I	IXXT	HQ
6609 (5698)	I	IXXT	HQ				

AN2D (RMBT)　　　　MINIATURE BUFFET CAR

Dia. AN207. Mark 2D. Converted from TSOT by the removal of another seating bay and fitting a proper buffet counter with boiler and microwave oven. Air conditioned. Stones equipment. 46S 1L. B4 bogies. ae. pa. q. ETH 5.

Lot No. 30822 Derby 1971. 33 t.

Note: Original numbers shown in parentheses. These vehicles carried 6602/10 – 2/5 when they were TSOTs.

6652 (5622)	I	ICCX	DY	6662 (5641)	I	ICCX	DY
6660 (5627)	I	ICCX	DY	6665 (5721)	I	ICCX	DY
6661 (5736)	I	ICCX	DY				

AN1F (RLO)　　　　SLEEPER RECEPTION CAR

Dia. AN101 (AN102*). Mark 2F. Converted from FO, these vehicles consist of pantry, microwave cooking facilities, seating area for passengers, telephone booth and staff toilet. 6703 – 8 also have a bar. Converted at RTC, Derby (6700), Ilford (6701 – 5) and Derby (6706 – 8). Air conditioned. ae. pa. q. B4 bogies. Fitted with payphone. 26F 1L.

6701 – 2/4/8. Lot No. 30859 Derby 1973 – 4. 33.5 t.
6703/5 – 7. Lot No. 30845 Derby 1973. 33.5 t.

Note: 6705 – 7 were also numbered 6430/21/18 when declassified prior to conversion.

6700 (3347)	I	IWCX	WB		6705 (3310)	I	IWCX	WB
6701 (3346)	*I	IWCX	WB		6706 (3283)	I	IWCX	WB
6702 (3421)	*I	IWCX	WB		6707 (3276)	I	IWCX	WB
6703 (3308)	I	IWCX	WB		6708 (3370)	I	IWCX	WB
6704 (3341)	I	IWCX	WB					

AC2F (TSO)   OPEN STANDARD

Dia. AC224. Mark 2F. Renumbered 1985 – 6 from FO. Converted 1990 to TSO with mainly unidirectional seating and power-operated sliding doors. Air conditioned. B4 bogies. 74S 2L + one tip-up seat. 6800 – 14 were converted by BREL Derby and have Temperature Ltd. air conditioning. 6815 – 29 were converted by RFS Industries Doncaster and have Stones air conditioning. ae. pa. ETH 5X. The coaches were also numbered 6435/42/39/43/9/36/40/52/4/5/1/37/48/63/65 and 6420/61/31/27/46/34/58/47/24/29/60/25/8/64/23 respectively.

6800 – 07. 6810 – 12. 6813 – 14. 6819/22/28. Lot No. 30859 Derby 1973 – 4. 33 t.
6808 – 6809. Lot No. 30873 Derby 1974 – 5. 33.5 t.
6815 – 18. 6820 – 21. 6823 – 27. 6829. Lot No. 30845 Derby 1973. 33 t.

6800 (3323)	I	IANR	NC		6815 (3282)	I	IANR	NC
6801 (3349)	I	IANR	NC		6816 (3316)	I	IANR	NC
6802 (3339)	I	IANR	NC		6817 (3311)	I	IANR	NC
6803 (3355)	I	IANR	NC		6818 (3298)	I	IANR	NC
6804 (3396)	I	IANR	NC		6819 (3365)	I	IANR	NC
6805 (3324)	I	IANR	NC		6820 (3320)	I	IANR	NC
6806 (3342)	I	IANR	NC		6821 (3281)	I	IANR	NC
6807 (3423)	I	IANR	NC		6822 (3376)	I	IANR	NC
6808 (3430)	I	IANR	NC		6823 (3289)	I	IANR	NC
6809 (3435)	I	IANR	NC		6824 (3307)	I	IANR	NC
6810 (3404)	I	IANR	NC		6825 (3301)	I	IANR	NC
6811 (3327)	I	IANR	NC		6826 (3294)	I	IANR	NC
6812 (3394)	I	IANR	NC		6827 (3306)	I	IANR	NC
6813 (3410)	I	IANR	NC		6828 (3380)	I	IANR	NC
6814 (3422)	I	IANR	NC		6829 (3288)	I	IANR	NC

NM5 (BSK)   INTER-CITY SANDITE COACH

Dia. NM503. Mark 2D. Air conditioned (Stones equipment). BFKs converted for use as Sandite coaches. Don't ask me why these are numbered as passenger coaches! 24F 1L. B4 Bogies. ETH 4.

Lot No. 30823 Derby 1971 – 2. 33.5 t.

6900	(17145) pg	I	IWCX	LL		6901	(17142) pg	I	IWCX	WB

AH2Z (BSOT)   OPEN BRAKE STANDARD (MICRO-BUFFET)

Dia. AH203. Mark 2. Converted from BSO by removal of one seating bay and replacing this by a counter with a space for a trolley. Adjacent toilet removed and converted to a steward's washing area/store. 23S 0L. ETH 4.

Lot No. 30757 Derby 1966.

9100 (9405)	**RR**	RAIS	IS	9105 (9404) pt	**RR**	RAIS	IS
9101 (9398)	**RR**	RAIS	IS				

AE2Z (BSO)   OPEN BRAKE STANDARD

Dia. AE203. Mark 2. These vehicles use the same body shell as the mark 2 BFK and have first class seat spacing and wider tables. Pressure ventilated. 31S 1L. B4 bogies. vd. pt. ETH 4.

Lot No. 30757 Derby 1966. 31.5 t.

9385	E	RAXX	IS	9414	E	RAIS	IS
9388	E	RAXX	IS				

Named vehicles:

9385 BALMACARA	9414 BRAHAN SEER
9388 BAILECHAUL	

AE2A (BSO)   OPEN BRAKE STANDARD

Dia. AE204. Mark 2A. These vehicles use the same body shell as the mark 2A BFK and have first class seat spacing and wider tables. Pressure ventilated. 31S 1L. B4 bogies. ad. ETH 4.

9417 – 9424. Lot No. 30777 Derby 1967. 31.5 t.
9428 – 9438. Lot No. 30788 Derby 1968. 31.5 t.

9417	**RR**	RCLL	LL	9428 pt		RCLL	LL
9418 pt	**RR**	RCLL	LL	9431 pt	**RR**	RCLL	LL
9419	**RR**	RCLL	LL	9434 pt	**RR**	RCLL	LL
9421	**RR**	RCLL	LL	9435	**RR**	RCLL	LL
9424 pt	**RR**	RCLL	LL	9438	I	RCLL	LL

AE2C (BSO)   OPEN BRAKE STANDARD

Dia. AE205. Mark 2C. Pressure ventilated. 31S 1L. B4 bogies. ad. ETH 4.

Lot No. 30820 Derby 1970. 32 t.

9458	**RR**	RCLL	LL

AE2D (BSO)   OPEN BRAKE STANDARD

Dia. AE206. Mark 2D. Air conditioned (Stones). 31S 1L. B4 bogies. ae. pg. ETH 5.

Lot No. 30824 Derby 1971. 33 t.

9479	I	ICCX	PC	9488	I	ICCX	PC
9480	I	ICCX	PC	9489	I	ICCX	PC
9481	I	ICCX	PC	9490	I	ICCX	MA
9482	I	IXXB	HQ	9492	I	IWRX	LA
9483	I	ILAG	DY	9493	I	ICCX	PC
9484	I	ICCX	PC	9494	I	ICCX	MA
9485	I	ICCX	PC	9495	I	ILAG	DY
9486	I	IWRX	LA				

AE2E (BSO)   OPEN BRAKE STANDARD

Dia. AE207. Mark 2E. Air conditioned (Stones). 32S 1L. B4 bogies. ae. pg. ETH 5.

Lot No. 30838 Derby 1972. 33 t.

9496	I	ICCX	PC	9503	I	ICCX	PC
9497	I	ICCX	PC	9504	I	ICCX	PC
9498	I	ICCX	PC	9505	I	ICCX	MA
9499	I	ILAG	DY	9506	I	ICCX	MA
9500	I	ICCX	MA	9507	I	ICCX	PC
9501	I	IWRX	LA	9508	I	ICCX	MA
9502	I	ICCX	MA	9509	I	ICCX	MA

AE2F (BSO)   OPEN BRAKE STANDARD

Dia. AE208. Mark 2F. Air conditioned (Temperature Ltd.). 32S 1L. B4 bogies. ae. pg. ETH 5X. All now r efurbished with power-operated vestibule doors, new panels and seat trim.

Lot No. 30861 Derby 1974. 34 t.

9513	I	ICCR	DY	9526 n	I	ICCR	DY
9516 n	I	ICCR	MA	9527	I	ICCR	MA
9520 n	I	ICCR	MA	9529	I	ICCR	MA
9521	I	ICCR	DY	9531	I	ICCR	MA
9522	I	ICCR	MA	9537 n	I	ICCR	DY
9523	I	ICCR	MA	9538	I	ICCR	MA
9524 n	I	ICCR	DY	9539	I	ICCR	DY
9525	I	ICCR	MA				

AF2F (DBSO)   DRIVING OPEN BRAKE STANDARD

Dia. AF201. Mark 2F. Air conditioned (Temperature Ltd.). Push & pull (t.d.m. system). Converted from BSO, these vehicles originally had half cabs at the brake end. They have since been refurbished and have had their cabs widened and the outer gangways removed. Fitted with cowcatchers. 32S 0L. B4 bogies. ae. pg. Cab to shore communication. BR Cellnet phone and data transmitter. ETH 5X.

9701 – 9710. Lot No. 30861 Derby 1974. Converted 1979. Disc brakes. 34 t.
9711 – 9713. Lot No. 30861 Derby 1974. Converted Glasgow 1985. 34 t.
9714. Lot No. 30861 Derby 1974. Converted Glasgow 1986. Disc brakes. 34 t.

9701 (9528)	I	IANR	NC	9709 (9515)	I	IANR	NC
9702 (9510)	I	IANR	NC	9710 (9518)	I	IANR	NC
9703 (9517)	I	IANR	NC	9711 (9532)	I	IANR	NC
9704 (9512)	I	IANR	NC	9712 (9534)	I	IANR	NC
9705 (9519)	I	IANR	NC	9713 (9535)	I	IANR	NC
9707 (9511)	I	IANR	NC	9714 (9536)	I	IANR	NC
9708 (9530)	I	IANR	NC				

AJ1G (RFM)   RESTAURANT BUFFET FIRST (MODULAR)

Dia. AJ103 (10200/1 are Dia. AJ101). Mark 3A. Air conditioned. Converted from HST TRFKs, RFBs and FOs. 22F (24F*). BT10 bogies. ae. pt. q. Fitted with payphone. ETH 14X.

10200 – 10211. Lot No. 30884 Derby 1977.
10212 – 10229. Lot No. 30878 Derby 1975 – 6. 39.80 t.
10230 – 10260. Lot No. 30890 Derby 1979. 39.80 t.

10200 (40519) *	I	IWCX	MA	10229 (11059)	I	IWCX	OY
10201 (40520) *	I	IWCX	MA	10230 (10021)	I	IWCX	PC
10202 (40504)	I	IWCX	MA	10231 (10016)	I	IWCX	PC
10203 (40506)	I	IANX	NC	10232 (10027)	I	IWCX	OY
10204 (40502)	I	IWCX	MA	10233 (10013)	I	IWCX	MA
10205 (40503)	I	IWCX	OY	10234 (10004)	I	IWCX	WB
10206 (40507)	I	IWCX	OY	10235 (10015)	I	IWCX	OY
10207 (40516)	I	IWCX	WB	10236 (10018)	I	IWCX	PC
10208 (40517)	I	IWCX	WB	10237 (10022)	I	IWCX	WB
10209 (40508)	I	IWCX	WB	10238 (10017)	I	IWCX	OY
10210 (40509)	I	IWCX	WB	10240 (10003)	I	IWCX	OY
10211 (40510)	I	IWCX	PC	10241 (10009)	I	IWCX	WB
10212 (11049)	I	IWCX	PC	10242 (10002)	I	IWCX	WB
10213 (11050)	I	IWCX	MA	10245 (10019)	I	IWCX	WB
10214 (11034)	I	IANX	NC	10246 (10014)	I	IWCX	OY
10215 (11032)	I	IWCX	WB	10247 (10011)	I	IANX	NC
10216 (11041)	I	IANX	NC	10248 (10005)	I	IWCX	WB
10217 (11051)	I	IWCX	OY	10249 (10012)	I	IWCX	MA
10218 (11053)	I	IWCX	MA	10250 (10020)	I	IWCX	OY
10219 (11047)	I	IWCX	OY	10251 (10024)	I	IWCX	OY
10220 (11056)	I	IWCX	OY	10252 (10008)	I	IWCX	WB
10221 (11012)	I	IWCX	PC	10253 (10026)	I	IWCX	WB
10222 (11063)	I	IWCX	MA	10254 (10006)	I	IWCX	MA
10223 (11043) s	I	IANX	NC	10255 (10010)	I	IWCX	OY
10224 (11062)	I	IWCX	MA	10256 (10028)	I	IWCX	PC
10225 (11014)	I	IWCX	OY	10257 (10007)	I	IWCX	MA
10226 (11015)	I	IWCX	MA	10258 (10023)	I	IWCX	WB
10227 (11057)	I	IWCX	MA	10259 (10025)	I	IWCX	OY
10228 (11035)	I	IANX	NC	10260 (10001)	I	IWCX	WB

AJ1J (RFM)   RESTAURANT BUFFET FIRST (MODULAR)

Dia. AJ105. Mark 4. Air conditioned. 20F 1L. SIG bogies (BT41). ae. pt. ETH   X.

Lot No. 31045 Metro-Cammell 1989 onwards. 45.5 t.

10300	I	IECX	BN	10317	I	IECX	BN
10301	I	IECX	BN	10318	I	IECX	BN
10302	I	IECX	BN	10319	I	IECX	BN
10303	I	IECX	BN	10320	I	IECX	BN
10304	I	IECX	BN	10321	I	IECX	BN
10305	I	IECX	BN	10322	I	IECX	BN
10306	I	IECX	BN	10323	I	IECX	BN
10307	I	IECX	BN	10324	I	IECX	BN
10308	I	IECX	BN	10325	I	IECX	BN
10309	I	IECX	BN	10326	I	IECX	BN
10310	I	IECX	BN	10327	I	IECX	BN
10311	I	IECX	BN	10328	I	IECX	BN
10312	I	IECX	BN	10329	I	IECX	BN
10313	I	IECX	BN	10330	I	IECX	BN
10314	I	IECX	BN	10331	I	IECX	BN
10315	I	IECX	BN	10332	I	IECX	BN
10316	I	IECX	BN	10333	I	IECX	BN

AU4G (SLEP)   SLEEPING CAR WITH PANTRY

Dia. AU401. Mark 3A. Air conditioned. 12 compartments with a fixed lower berth and a hinged upper berth, plus an attendants compartment with 2L (controlled emission). BT10 bogies. ae. ETH 7X.

Lot No. 30960 Derby 1981 – 3.

10500	I	ICHV	BN	10534	I	IWRR	LA
10501	I	IWCR	WB	10535	I	ICCR	PC
10502	I	IWCR	WB	10536	I	IWCR	WB
10503	I	ICHV	BN	10537	I	ICCR	PC
10504	I	IWCR	WB	10538	I	IWCX	WB
10506	I	IWCR	WB	10539	I	IWRX	LA
10507	I	IWCR	WB	10540	I	IWRX	LA
10508	I	IWCR	WB	10541	I	IWCX	WB
10510	I	IWCR	WB	10542	I	IWCR	WB
10512	I	IWCR	WB	10543	I	IWCR	WB
10513	I	IWCR	WB	10544	I	IWCR	WB
10514	I	ICHV	BN	10546	I	ICCR	PC
10515	I	IWCR	WB	10547	I	IWCR	WB
10516	I	IWCR	WB	10548	I	IWCR	WB
10519	I	IWCR	WB	10549	I	ICCX	PC
10520	I	IWCR	WB	10550	I	IXXH	HQ
10522	I	IWCR	WB	10551	I	IWCR	WB
10523	I	IWCR	WB	10553	I	IWCR	WB
10526	I	IWCR	WB	10554	I	ICCR	PC
10527	I	IWCR	WB	10555	I	IWCR	WB
10529	I	IWCR	WB	10556	I	IWRX	LA
10530	I	ICCR	PC	10557	I	ICHV	BN
10531	I	ICCR	PC	10558	I	IWCX	WB
10532	I	ICCR	PC	10559	I	IWCR	WB
10533	I	ICHV	BN	10560	I	IWCX	WB

10561	I	IWCR	WB	10591	I	IXXH	HQ
10562	I	IWCR	WB	10592	I	IXXH	HQ
10563	I	ICCR	PC	10593	I	IWCX	WB
10565	I	IWCR	WB	10594	I	ICCR	PC
10566	I	ICCX	PC	10595		IXXH	HQ
10567	I	IXXH	HQ	10596	I	IWRR	LA
10569	I	IWCX	WB	10597	I	IWCR	WB
10570	I	IXXH	HQ	10598	I	IWCR	WB
10571	I	ICHV	BN	10599	I	IXXH	HQ
10572	I	ICCR	PC	10600	I	IWCR	WB
10573	I	IWCX	WB	10601	I	ICCR	PC
10574	I	ICHV	BN	10602	I	ICCX	PC
10575	I	ICHV	BN	10603	I	IXXH	HQ
10577		IXXH	HQ	10604	I	IWCX	WB
10578	I	IXXH	HQ	10605	I	IWRR	LA
10579		IXXH	HQ	10606	I	IXXH	HQ
10580	I	ICCR	PC	10607	I	IWCR	WB
10581		IXXH	HQ	10609		IXXH	HQ
10582	I	ICHV	BN	10610	I	IWCR	WB
10583	I	IWRR	LA	10612	I	IWRR	LA
10584	I	IWRR	LA	10613	I	IWCR	WB
10586	I	IWCX	WB	10614	I	IWCR	WB
10588	I	IWRR	LA	10616	I	IWRR	LA
10589	I	IWCR	WB	10617	I	IWCR	WB
10590	I	IWCR	WB				

AS4G (SLE)   SLEEPING CAR

Dia. AS403. Mark 3A. Air conditioned. 13 compartments with a fixed lower berth and a hinged upper berth. 2L (controlled emission). BT10 bogies. ae. ETH 6X.

Lot No. 30961 Derby 1980 – 4.

10646	I	ICHV	BN	10670	I	IXXH	HQ
10647	I	IWCR	WB	10672	I	IWCR	WB
10648	I	IWCR	WB	10673		IXXH	HQ
10649	I	IWCR	WB	10674	I	IWCR	WB
10650	I	IWCR	WB	10675	I	IWCR	WB
10651	I	IWCR	WB	10678		IXXH	HQ
10653	I	IWCR	WB	10679		IXXH	HQ
10654	I	IWCR	WB	10680	I	IWCR	WB
10655	I	ICHV	BN	10682	I	IWCR	WB
10656	I	IXXH	HQ	10683	I	IWCR	WB
10657	I	ICHV	BN	10684		IXXH	HQ
10658	I	IWCR	WB	10685	I	IWCR	WB
10660	I	IWCR	WB	10686	I	IWCR	WB
10661	I	IXXH	HQ	10687	I	IWCR	WB
10662	I	IXXH	HQ	10688	I	IWCR	WB
10663	I	IWCR	WB	10689	I	IWCR	WB
10665	I	IXXH	HQ	10690	I	IWCR	WB
10666	I	IWCR	WB	10691	I	IWCR	WB
10668	I	IWCR	WB	10692	I	IWCR	WB

10693	I	IWCR	WB	10714	I	IWCR	WB
10696	I	IWCR	WB	10715	I	IXXC	HQ
10697	I	IWCR	WB	10716	I	IWCR	WB
10699	I	IWCR	WB	10717	I	IWCR	WB
10700		IXXH	HQ	10718	I	IWCR	WB
10701	I	IWCR	WB	10719	I	IWCR	WB
10702	I	ICHV	BN	10720	I	IXXH	HQ
10703	I	IWCR	WB	10722	I	IWCR	WB
10704	I	IXXC	HQ	10723	I	IWCR	WB
10705	I	IXXH	HQ	10724	I	ICHV	BN
10706	I	IWCR	WB	10725	I	ICHV	BN
10707	I	IXXH	HQ	10726	I	ICHV	BN
10708	I	IWCR	WB	10727	I	ICHV	BN
10709	I	IWCR	WB	10728	I	IXXH	HQ
10710	I	IWCR	WB	10729	I	ICHV	BN
10711	I	IWCR	WB	10730	I	IWCR	WB
10712	I	IWCR	WB	10731	I	IWCR	WB
10713	I	IXXC	HQ	10732	I	IWCR	WB

AD1G (FO)   OPEN FIRST

Dia. AD108. Mark 3A. Air conditioned. 48F 2L. BT10 bogies (BT15 b). ae. pa.
ETH 6X. All now facelifted with new upholstery, carpets etc. 11005 – 7 have
regained their original numbers, having being converted back from open composites 11905 – 7.

Lot No. 30878 Derby 1975 – 6. 34.30 t.

11005	I	IWCX	WB	11031	I	IWCX	MA
11006	I	IWCX	WB	11033	I	IWCX	WB
11007	I	IWCX	MA	11036	I	IWCX	WB
11011 z	I	IWCX	PC	11037	I	IWCX	PC
11013	I	IWCX	WB	11038	I	IWCX	PC
11016	I	IWCX	PC	11039	I	IWCX	PC
11017	I	IWCX	PC	11040	I	IWCX	PC
11018	I	IWCX	WB	11042	I	IWCX	WB
11019	I	IWCX	PC	11044	I	IWCX	MA
11020	I	IWCX	WB	11045	I	IWCX	PC
11021 b	I	IWCX	WB	11046	I	IWCX	WB
11023	I	IWCX	PC	11048	I	IWCX	PC
11024	I	IWCX	WB	11052	I	IWCX	WB
11026	I	IWCX	MA	11054	I	IWCX	WB
11027	I	IWCX	PC	11055	I	IWCX	PC
11028	I	IWCX	WB	11058	I	IWCX	WB
11029	I	IWCX	MA	11060	I	IWCX	WB
11030	I	IWCX	WB				

AD1H (FO)   OPEN FIRST

Dia. AD109. Mark 3B. Air conditioned. 48F 2L. BT10 bogies. ae. pa. ETH 6X.
Inter-City 80 seats. Some of these coaches were named, but the names are
now being removed.

Lot No. 30982 Derby 1985. 36.46 t.

11064	I	IWCX	MA	11083 p	I	IWCX	WB
11065	I	IWCX	MA	11084 p	I	IWCX	WB
11066	I	IWCX	MA	11085 p	I	IWCX	MA
11067	I	IWCX	WB	11086 p	I	IWCX	WB
11068	I	IWCX	WB	11087 p	I	IWCX	MA
11069	I	IWCX	MA	11088 p	I	IWCX	WB
11070	I	IWCX	MA	11089 p	I	IWCX	MA
11071	I	IWCX	MA	11090 p	I	IWCX	WB
11072	I	IWCX	WB	11091 p	I	IWCX	MA
11073	I	IWCX	MA	11092 p	I	IWCX	WB
11074	I	IWCX	WB	11093 p	I	IWCX	MA
11075	I	IWCX	MA	11094 p	I	IWCX	MA
11076	I	IWCX	WB	11095 p	I	IWCX	MA
11077	I	IWCX	WB	11096 p	I	IWCX	MA
11078	I	IWCX	MA	11097 p	I	IWCX	WB
11079	I	IWCX	WB	11098 p	I	IWCX	WB
11080	I	IWCX	WB	11099 p	I	IWCX	WB
11081	I	IWCX	WB	11100 p	I	IWCX	MA
11082	I	IWCX	WB	11101 p	I	IWCX	MA

AD1J (FO)   OPEN FIRST

Dia. AD111. Mark 4. Air conditioned. Known as 'Pullman open' by BR. 46F 1L. SIG bogies (BT41). ae. pa. ETH 6.

Note: 11264 – 71 have now been cancelled.

Lot No. 31046 Metro-Cammell 1989 onwards. 39.70 t.

11200	I	IECX	BN	11223	I	IECX	BN
11201 p	I	IECX	BN	11224	I	IECX	BN
11202	I	IECX	BN	11225 p	I	IECX	BN
11203 p	I	IECX	BN	11226	I	IECX	BN
11204 p	I	IECX	BN	11227 p	I	IECX	BN
11205	I	IECX	BN	11228 p	I	IECX	BN
11206	I	IECX	BN	11229 p	I	IECX	BN
11207 p	I	IECX	BN	11230	I	IECX	BN
11208	I	IECX	BN	11231 p	I	IECX	BN
11209	I	IECX	BN	11232	I	IECX	BN
11210	I	IECX	BN	11233 p	I	IECX	BN
11211 p	I	IECX	BN	11234	I	IECX	BN
11212	I	IECX	BN	11235 p	I	IECX	BN
11213 p	I	IECX	BN	11236	I	IECX	BN
11214 p	I	IECX	BN	11237 p	I	IECX	BN
11215	I	IECX	BN	11238	I	IECX	BN
11216	I	IECX	BN	11239 p	I	IECX	BN
11217 p	I	IECX	BN	11240	I	IECX	BN
11218	I	IECX	BN	11241	I	IECX	BN
11219 p	I	IECX	BN	11242 p	I	IECX	BN
11220	I	IECX	BN	11243 p	I	IECX	BN
11221 p	I	IECX	BN	11244	I	IECX	BN
11222 p	I	IECX	BN	11245 p	I	IECX	BN

11246 p	I	IECX	BN
11247 p	I	IECX	BN
11248	I	IECX	BN
11249 p	I	IECX	BN
11250	I	IECX	BN
11251 p	I	IECX	BN
11252	I	IECX	BN
11253 p	I	IECX	BN
11254	I	IECX	BN
11255 p	I	IECX	BN
11256	I	IECX	BN
11257 p	I	IECX	BN

11258	I	IECX	BN
11259 p	I	IECX	BN
11260	I	IECX	BN
11261 p	I	IECX	BN
11262	I	IECX	BN
11263 p	I	IECX	BN
11272	I	IECX	BN
11273	I	IECX	BN
11274	I	IECX	BN
11275	I	IECX	BN
11276	I	IECX	BN

AC2G (TSO)   OPEN STANDARD

Dia. AC213 (AC220 z). Mark 3A. Air conditioned. All now refurbished with modified seat backs and new layout. 76S 2L (74S 2L z). BT10 (BREL T4*, BT15 b) bogies. ae. pa. ETH 6X. 12169 – 72 have been converted from open composites 11908 – 10/22, formerly FOs 11008 – 10/22.

Lot No. 30877 Derby 1975 – 7. 34.30 t.

12004	I	IWCX	MA
12005	I	IWCX	WB
12007	I	IWCX	WB
12008	I	IWCX	WB
12009	I	IWCX	MA
12010 b	I	IWCX	WB
12011	I	IWCX	WB
12012	I	IWCX	WB
12013	I	IWCX	WB
12014	I	IWCX	WB
12015	I	IWCX	WB
12016	I	IWCX	WB
12017	I	IWCX	WB
12019	I	IWCX	WB
12020	I	IWCX	WB
12021	I	IWCX	WB
12022	I	IWCX	WB
12023	I	IWCX	WB
12024	I	IWCX	MA
12025	I	IWCX	WB
12026 w	I	IWCX	MA
12027	I	IWCX	WB
12028	I	IWCX	WB
12029	I	IWCX	PC
12030	I	IWCX	WB
12031	I	IWCX	MA
12032	I	IWCX	PC
12033 z	I	IWCX	WB
12034	I	IWCX	MA
12035	I	IWCX	MA
12036 s	I	IWCX	MA

12037	I	IWCX	MA
12038	I	IWCX	MA
12040	I	IWCX	PC
12041	I	IWCX	PC
12042 w	I	IWCX	WB
12043	I	IWCX	MA
12044	I	IWCX	MA
12045	I	IWCX	WB
12046	I	IWCX	PC
12047 z	I	IWCX	PC
12048	I	IWCX	WB
12049	I	IWCX	PC
12050 w	I	IWCX	MA
12051	I	IWCX	MA
12052	I	IWCX	WB
12053	I	IWCX	MA
12054 z	I	IWCX	MA
12055	I	IWCX	MA
12056	I	IWCX	MA
12057	I	IWCX	WB
12058	I	IWCX	WB
12059 w	I	IWCX	WB
12060	I	IWCX	MA
12061 w	I	IWCX	WB
12062	I	IWCX	MA
12063	I	IWCX	PC
12064	I	IWCX	PC
12065	I	IWCX	PC
12066	I	IWCX	WB
12067	I	IWCX	WB
12068	I	IWCX	WB

12069		I	IWCX	MA	12121	I	IWCX	WB

12069		I	IWCX	MA	12121		I	IWCX	WB
12070		I	IWCX	WB	12122 z		I	IWCX	WB
12071		I	IWCX	WB	12123		I	IWCX	WB
12072		I	IWCX	WB	12124		I	IWCX	MA
12073		I	IWCX	MA	12125		I	IWCX	MA
12075		I	IWCX	MA	12126		I	IWCX	MA
12076		I	IWCX	WB	12127		I	IWCX	WB
12077		I	IWCX	WB	12128 w		I	IWCX	MA
12078		I	IWCX	PC	12129		I	IWCX	MA
12079		I	IWCX	MA	12130		I	IWCX	MA
12080		I	IWCX	MA	12131		I	IWCX	MA
12081		I	IWCX	PC	12132		I	IWCX	MA
12082		I	IWCX	PC	12133		I	IWCX	PC
12083		I	IWCX	WB	12134		I	IWCX	WB
12084		I	IWCX	WB	12135		I	IWCX	PC
12085 w		I	IWCX	WB	12136		I	IWCX	WB
12086 w		I	IWCX	WB	12137		I	IWCX	PC
12087 w		I	IWCX	MA	12138		I	IWCX	WB
12088 z		I	IWCX	PC	12139		I	IWCX	MA
12089		I	IWCX	PC	12140 *z		I	IWCX	WB
12090		I	IWCX	WB	12141		I	IWCX	WB
12091		I	IWCX	WB	12142 z		I	IWCX	PC
12092		I	IWCX	MA	12143		I	IWCX	WB
12093		I	IWCX	WB	12144 z		I	IWCX	WB
12094		I	IWCX	WB	12145		I	IWCX	PC
12095		I	IWCX	PC	12146		I	IWCX	MA
12096		I	IWCX	PC	12147		I	IWCX	PC
12097		I	IWCX	PC	12148		I	IWCX	PC
12098		I	IWCX	WB	12149		I	IWCX	MA
12099		I	IWCX	PC	12150		I	IWCX	MA
12100 z		I	IWCX	PC	12151		I	IWCX	PC
12101 w		I	IWCX	WB	12152		I	IWCX	WB
12102		I	IWCX	WB	12153		I	IWCX	PC
12103 z		I	IWCX	PC	12154		I	IWCX	MA
12104		I	IWCX	PC	12155		I	IWCX	WB
12105		I	IWCX	MA	12156		I	IWCX	WB
12106		I	IWCX	WB	12157		I	IWCX	MA
12107		I	IWCX	MA	12158		I	IWCX	WB
12108 w		I	IWCX	MA	12159		I	IWCX	MA
12109 w		I	IWCX	MA	12160 w		I	IWCX	MA
12110		I	IWCX	MA	12161 z		I	IWCX	PC
12111		I	IWCX	WB	12163		I	IWCX	WB
12112 z		I	IWCX	MA	12164		I	IWCX	MA
12113		I	IWCX	WB	12165		I	IWCX	PC
12114		I	IWCX	MA	12166		I	IWCX	PC
12115		I	IWCX	PC	12167		I	IWCX	MA
12116		I	IWCX	PC	12168 w		I	IWCX	PC
12117		I	IWCX	MA	12169 z		I	IWCX	WB
12118		I	IWCX	WB	12170 z		I	IWCX	WB
12119		I	IWCX	PC	12171 z		I	IWCX	WB
12120		I	IWCX	MA	12172 z		I	IWCX	WB

AI2J (TSOE)   OPEN STANDARD (END)

Dia. AI201. Mark 4. Air conditioned. 74S 2L. SIG bogies (BT41). ae. pa. ETH 6.

Note: 12232 was converted from the original 12405.

Lot No. 31047 Metro-Cammell 1989 onwards. 39.5 t.

12200	I	IECX	BN	12216	I	IECX	BN
12201	I	IECX	BN	12217	I	IECX	BN
12202	I	IECX	BN	12218	I	IECX	BN
12203	I	IECX	BN	12219	I	IECX	BN
12204	I	IECX	BN	12220	I	IECX	BN
12205	I	IECX	BN	12222	I	IECX	BN
12206	I	IECX	BN	12223	I	IECX	BN
12207	I	IECX	BN	12224	I	IECX	BN
12208	I	IECX	BN	12225	I	IECX	BN
12209	I	IECX	BN	12226	I	IECX	BN
12210	I	IECX	BN	12227	I	IECX	BN
12211	I	IECX	BN	12228	I	IECX	BN
12212	I	IECX	BN	12229	I	IECX	BN
12213	I	IECX	BN	12230	I	IECX	BN
12214	I	IECX	BN	12231	I	IECX	BN
12215	I	IECX	BN	12232	I	IECX	BN

AL2J (TSOD) OPEN STANDARD (DISABLED ACCESS)

Dia. AL201. Mark 4. Air conditioned. 72S + wheelchair space 1L (suitable for a disabled person). SIG bogies (BT41). ae. pa. p. ETH 6.

Lot No. 31048 Metro-Cammell 1989 onwards. 39.4 t.

12300	I	IECX	BN	12316	I	IECX	BN
12301	I	IECX	BN	12317	I	IECX	BN
12302	I	IECX	BN	12318	I	IECX	BN
12303	I	IECX	BN	12319	I	IECX	BN
12304	I	IECX	BN	12320	I	IECX	BN
12305	I	IECX	BN	12321	I	IECX	BN
12306	I	IECX	BN	12322	I	IECX	BN
12307	I	IECX	BN	12323	I	IECX	BN
12308	I	IECX	BN	12324	I	IECX	BN
12309	I	IECX	BN	12325	I	IECX	BN
12310	I	IECX	BN	12326	I	IECX	BN
12311	I	IECX	BN	12327	I	IECX	BN
12312	I	IECX	BN	12328	I	IECX	BN
12313	I	IECX	BN	12329	I	IECX	BN
12314	I	IECX	BN	12330	I	IECX	BN
12315	I	IECX	BN				

AC2J (TSO)   OPEN STANDARD

Dia. AC214. Mark 4. Air conditioned. 74S 2L. SIG bogies (BT41). ae. pa.
ETH 6X.

Notes: 12405 is the second coach to carry that number. It was built from the
bodyshell originally intended for 12221. The original 12405 is now 12232.
12490 – 12512 were cancelled.

Lot No. 31049 Metro-Cammell 1989 onwards. 39.9 t.

12400	I	IECX	BN	12441	I	IECX	BN
12401	I	IECX	BN	12442	I	IECX	BN
12402	I	IECX	BN	12443	I	IECX	BN
12403	I	IECX	BN	12444	I	IECX	BN
12404	I	IECX	BN	12445	I	IECX	BN
12405	I	IECX	BN	12446	I	IECX	BN
12406	I	IECX	BN	12447	I	IECX	BN
12407	I	IECX	BN	12448	I	IECX	BN
12408	I	IECX	BN	12449	I	IECX	BN
12409	I	IECX	BN	12450	I	IECX	BN
12410	I	IECX	BN	12451	I	IECX	BN
12411	I	IECX	BN	12452	I	IECX	BN
12412	I	IECX	BN	12453	I	IECX	BN
12413	I	IECX	BN	12454	I	IECX	BN
12414	I	IECX	BN	12455	I	IECX	BN
12415	I	IECX	BN	12456	I	IECX	BN
12416	I	IECX	BN	12457	I	IECX	BN
12417	I	IECX	BN	12458	I	IECX	BN
12418	I	IECX	BN	12459	I	IECX	BN
12419	I	IECX	BN	12460	I	IECX	BN
12420	I	IECX	BN	12461	I	IECX	BN
12421	I	IECX	BN	12462	I	IECX	BN
12422	I	IECX	BN	12463	I	IECX	BN
12423	I	IECX	BN	12464	I	IECX	BN
12424	I	IECX	BN	12465	I	IECX	BN
12425	I	IECX	BN	12466	I	IECX	BN
12426	I	IECX	BN	12467	I	IECX	BN
12427	I	IECX	BN	12468	I	IECX	BN
12428	I	IECX	BN	12469	I	IECX	BN
12429	I	IECX	BN	12470	I	IECX	BN
12430	I	IECX	BN	12471	I	IECX	BN
12431	I	IECX	BN	12472	I	IECX	BN
12432	I	IECX	BN	12473	I	IECX	BN
12433	I	IECX	BN	12474	I	IECX	BN
12434	I	IECX	BN	12475	I	IECX	BN
12435	I	IECX	BN	12476	I	IECX	BN
12436	I	IECX	BN	12477	I	IECX	BN
12437	I	IECX	BN	12478	I	IECX	BN
12438	I	IECX	BN	12479	I	IECX	BN
12439	I	IECX	BN	12480	I	IECX	BN
12440	I	IECX	BN	12481	I	IECX	BN

12482	I IECX	BN	12522	I IECX	BN
12483	I IECX	BN	12523	I IECX	BN
12484	I IECX	BN	12524	I IECX	BN
12485	I IECX	BN	12525	I IECX	BN
12486	I IECX	BN	12526	I IECX	BN
12487	I IECX	BN	12527	I IECX	BN
12488	I IECX	BN	12528	I IECX	BN
12489	I IECX	BN	12529	I IECX	BN
12513	I IECX	BN	12530	I IECX	BN
12514	I IECX	BN	12531	I IECX	BN
12515	I IECX	BN	12532	I IECX	BN
12516	I IECX	BN	12533	I IECX	BN
12517	I IECX	BN	12534	I IECX	BN
12518	I IECX	BN	12535	I IECX	BN
12519	I IECX	BN	12536	I IECX	BN
12520	I IECX	BN	12537	I IECX	BN
12521	I IECX	BN	12538	I IECX	BN

AA11 (FK)   CORRIDOR FIRST

Dia. AA101. Mark 1. 42F 2L. ETH 3. d.

Note: 13233/6/7, 13303/14/6/26/35 are on loan to the Humberside Loco Preservation Group at Hull Dairycoates (HU) and 13230 is on loan to the SRPS at Bo'ness.

13225 – 13227. Lot No. 30381 Ashford/Eastleigh 1959. B4 bogies. 33 t.
13306 – 13344. Lot No. 30667 Swindon 1962. Commonwealth bogies. 36 t.

13225	xk RR RCLL	LL	13318	a pa I ICHV	BN
13227	xk RR RCLL	LL	13341	vf I ICHV	BN
13306	v ICHH	CL	13344	vf ICHH	CL

AA1D (FK)   CORRIDOR FIRST

Dia. AA109. Mark 2D. Air conditioned (Stones). 42F 2L. B4 bogies. ae. pa. ETH 5. 13585 – 13607 require at least 800 V train heat supply.

Lot No. 30825 Derby 1971 – 2. 34.5 t.

13581	I IXXH	HQ	13595	I IXXH	HQ
13582	I ICHH	CL	13596	I IXXH	HQ
13583	I IXXH	HQ	13601	I IXXB	HQ
13585	I ICHH	CL	13603	I IXXB	HQ
13592	I IXXB	HQ	13604	I ICHV	BN
13593	I IXXH	HQ	13607	I ICHV	BN

SPECIAL NOTE: All BFKs were formerly numbered in the 14xxx series. Subtract 3000 from present number to obtain former number.

AB11 (BFK)   CORRIDOR BRAKE FIRST

Dia. AB101. Mark 1. 24F 1L. vd. ETH 2.

17015. Lot No. 30668 Swindon 1961. Commonwealth bogies. 36 t.
17023. Lot No. 30718 Swindon 1963. Commonwealth bogies and metal window frames. 36 t.

17015 x	I	ICHV	BN	17023 x	I	ICHV	BN

AB1A (BFK)   CORRIDOR BRAKE FIRST

Dia. AB103. Mark 2A. Pressure ventilated. 24F 1L. B4 bogies. d. ETH 4.

17064 – 17077. Lot No. 30775 Derby 1967 – 8. 32 t.
17086 – 17099. Lot No. 30786 Derby 1968. 32 t.

17089/90 were renumbered 35502/3 for a time when they were previously declassified.

17064 v		RR RAIS	IS	17089 v		RR RAIS	IS
17068 v		RR RAIS	IS	17090 v		RR RAXX	IS
17076 a pt	N	NWXX	EH	17091 v		RR RAIS	IS
17077 a	N	RBHT	HT	17096 a pt	N	NSSX	EH
17086 a pt	N	RBHT	HT	17099 v		RR RAIS	IS

AB1D (BFK)   CORRIDOR BRAKE FIRST

Dia. AB106. Mark 2D. Air conditioned (Stones equipment). 24F 1L. B4 Bogies.
17163 – 17172 require at least 800 V train heat supply. ae. pt. ETH 5.

Lot No. 30823 Derby 1971 – 2. 33.5 t.

17141 pg I	ICHV	BN	17161 pg I	ICHH	HQ
17144 I	ICHD	CL	17163 I	IXXT	HQ
17146 pg I	ICHX	CL	17164 pg I	ICHX	BN
17148 pg I	IXXT	HQ	17165 I	ICHD	CL
17151 pg I	ICHD	CL	17166 pg I	IWCD	CL
17153 pg I	ICHV	BN	17167 pg I	ICHD	CL
17155 I	IXXT	HQ	17169 pg I	ICHD	CL
17156 I	ICHD	CL	17170 pg I	ICHX	CL
17158 pg I	IXXB	HQ	17171 I	ICHH	HQ
17159 pg I	ICHD	CL	17172 pg I	ICHX	BN

AE1G (BFO)   OPEN BRAKE FIRST

Dia. AE101. Mark 3B. Air conditioned. 36F 1L. BT10 bogies. ae. pg. ETH 5X.
Fitted with hydraulic handbrake.

Lot No. 30990 Derby 1986. 35.81 t.

17173 I	IWCX	MA	17175 I	IWCX	MA
17174 I	IWCX	MA			

AB31 (BCK)   CORRIDOR BRAKE COMPOSITE

Dia.AB301 (AB302*). Mark 1. As with the CKs there are two variants depending upon whether the standard class compartments have armrests. Each vehicle has two first class and three standard class compartments. 12F 18S 2L (12F 24S 2L*). ETH 2.

21241 – 21246. Lot No. 30669 Swindon 1961 – 2. Commonwealth bogies. 36 t.
21265 – 21274. Lot No. 30732 Derby 1964. Commonwealth bogies. 37 t.

21241	vd	E	ICHS	CL	21268 *ae I	ICHV	BN
21246	xd	I	ICHV	BN	21269 *ad I	ICHV	BN
21265	*ae		ICHH	CL	21274 *ae I	ICHV	BN
21266	*ae	I	ICHV	BN			

AB21 (BSK)   CORRIDOR BRAKE STANDARD

Dia. AB201. Mark 1. Each vehicle has four compartments. All remaining vehicles have metal window frames and melamine interior panelling 24S 1L. ETH 2. vd.

35317. Lot No. 30699 Wolverton 1962 – 3. Commonwealth bogies. 37 t.
35452 – 35479. Lot No. 30721 Wolverton 1963. Commonwealth bogies. 37 t.

g – Converted to ETH generator vehicle for Flying Scotsman Services. Still in BR stock. Carries No. '196'.

35317	pt	NI ICHS	CL	35469 g	M ICHL	BN
35452	pt	NR RCXX	LL	35479 pt	NI ICHS	CL
35453		RR RCXX	LL			

AB2A (BSK)   CORRIDOR BRAKE STANDARD

Dia. AB204. Mark 2A. Pressure ventilated. Renumbered from BFK. 24S 1L. B4 bogies. d. ETH 4.

35500/15 – 18. Lot No. 30786 Derby 1968. 32 t.
35512 – 14. Lot No. 30796 Derby 1969 – 70. 32.5 t.

35500 (17094) v	RR RAXX	IS	35515 (17079) a	NR RCLL	LL
35512 (17057) a pt	RR RCLL	LL	35516 (17080) a pt	RR RCLL	LL
35513 (17063) a pt	NR RCLL	LL	35517 (17088) a pt	NR RCLL	LL
35514 (17069) a pt	NR RCLL	LL	35518 (17097) a pt	NR RCLL	LL

THE PLATFORM 5
TRANSPORT BOOK CLUB
MEMBERSHIP APPLICATION FORM

To enrol for one year's membership in the PLATFORM 5 TRANSPORT BOOK CLUB, please complete this form (or a photocopy) and send it with your cheque/postal order for £1.50 made payable to 'Platform 5 Publishing Limited' to:

The Platform 5 Transport Book Club, Wyvern House, Sark Road, SHEFFIELD, S2 4HG.

BLOCK CAPITALS PLEASE

Name: . . . .

Address: . . . .

. . . .

Post Code:. . . . .

Telephone: . . . .

Please accept my application and enrol me as a member of the Platform 5 Transport Book Club.

As a member I will receive four issues of the club newsletter, each containing a number of new books at prices of at least 15% less than the published cover price (exclusive of postage and packing).

I understand I am not obliged to buy any of the books offered, and there is no limit to the number of copies of each book that may be ordered.

I enclose my cheque/postal order for £1.50 payable to Platform 5 Publishing Limited.

Signed: . . . .

Date: . . . .

Office Use Only:. . . . .

PLATFORM 5 PUBLISHING LTD.
MAIL ORDER LIST

NEW TITLES   Price

BR Pocket Book No. 1: Locomotives	1.85
BR Pocket Book No. 2: Coaching Stock	1.85
BR Pocket Book No. 3: DMUs & Channel Tunnel Stock	1.85
BR Pocket Book No. 4: EMUs	1.85
British Railways Locomotives & Coaching Stock 1994 **MAR 94**	7.50
Light Rail Review 5	7.50
Preserved Locomotives of British Railways 8th edition	6.95
Exeter – Newton Abbot – A Railway History	25.00
Manx Electric	8.95
Steam Alive (Friends of the NRM)	2.95
Departmental Coaching Stock 5th edition (SCTP)	6.95
Buses in Britain (Capital)	19.95
Going Green (Capital)	5.95
TGV Handbook (Capital)	7.95

Modern British Railway Titles

On-Track Plant on British Railways 4th Edition	5.50
The Fifty 50s in Colour	5.95
Blood, Sweat and Fifties (Class 50 Society)	2.95
British Rail Internal Users (SCTP)	7.95
British Rail Wagon Fleet – Air Braked Stock (SCTP)	6.95
British Rail Wagon Fleet Volume 5 (SCTP)	5.25
RIV Wagon Fleet (SCTP)	5.95
British Rail Passenger Trains (Capital)	7.95
This is London Transport (Capital)	4.95
Miles & Chains Volume 2 – London Midland (Milepost)	1.40
Miles & Chains Vol. 3 – Scottish, Vol. 5 – Southern (Milepost)	each 1.00
Class Fifty Factfile (Class 50 Society)	2.95

Overseas Railways

German Railways Locomotives & MUs 3rd edition	12.50
Swiss Railways/Chemins de Fer Suisses	9.95
French Railways/Chemins de Fer Francais 2nd edition	9.95
ÖBB/Austrian Federal Railways 2nd edition	6.95
Benelux Locomotives & Coaching Stock 2nd edition	6.95
A Guide to Portuguese Railways (Fearless)	4.95
Paris Metro Handbook (Capital)	7.95
World Metro Systems (Capital)	6.95

Historical Railway Titles

6203 'Princess Margaret Rose'	19.95
Midland Railway Portrait	12.95
Steam Days on BR 1 – The Midland Line in Sheffield	4.95
Rails along the Sea Wall (Dawlish – Teignmouth Pictorial)	4.95
The Rolling Rivers	6.95
British Baltic Tanks	6.95
The Railways of Winchester	6.95
Register of Closed Railways 1948 – 91 (Milepost)	5.95
LNWR Branch Lines of West Leics & East Warwicks (Milepost)	7.95
British Railways Mark 1 Coaches (Atlantic)	19.95

Mallard – The Record Breaker (Friends of the NRM)	4.95
Rails Through The Clay (Capital)	25.00
The 1938 Tube Stock (Capital)	9.95
Metropolitan Steam Locomotives (Capital)	9.95
The First Tube (Capital)	4.95
Register of Closed Railways 1948 – 91 (Milepost)	5.95

Political

The Battle for the Settle & Carlisle	6.95

Rambling

Rambles by Rail 1 – The Hope Valley Line	1.95
Rambles by Rail 2 – Liskeard-Looe	1.95
Rambles by Rail 4 – The New Forest	1.95

Light Rail Transit, Trams & Buses

Light Rail Review 1 (Reprint)	6.95
Light Rail Review 2/3/4	each 7.50
UK Light Rail Systems No.1: Manchester Metrolink	8.50
Blackpool & Fleetwood By Tram	7.50
London Bus Handbook Part 1 (Capital)	8.95
Bus Handbook 9: Wales (Capital)	7.95
Bus Handbook 10: Scotland (Capital)	8.95
Routemaster Handbook (Capital)	7.95
Bus Review 8 (Bus Enthusiast)	4.95
Edinburgh's Trams & Buses (Bus Enthusiast)	4.95

Maps and Track Diagrams (Quail Map Company)

British Rail Track Diagrams 1 – Scotland & Isle of Man	5.00
British Rail Track Diagrams 3 – Western	5.00
British Rail Track Diagrams 4 – London Midland	6.95
London Railway Map	5.95
London Transport Track Map	1.30
China Railway Atlas	3.50
Czech Republic & Slovakia Railway Map	1.70
Greece Railway Map	1.00
Poland Railway Map	2.00

PVC Book Covers

A6 Pocket Book Covers in Blue, Red, Green or Grey	0.80
Locomotives & CS Covers in Blue, Red, Green or Grey	1.00
A5 Book Covers in Blue, Red, Green or Grey	1.40

Back Numbers

Locomotives & Coaching Stock 1985	2.95
Locomotives & Coaching Stock 1986	3.30
Locomotives & Coaching Stock 1987	3.30
Locomotives & Coaching Stock 1988	3.95
Locomotives & Coaching Stock 1989	4.95
Locomotives & Coaching Stock 1990	5.95
Locomotives & Coaching Stock 1991	6.60
British Railways Locomotives & Coaching Stock 1992	7.00
British Railways Locomotives & Coaching Stock 1993	7.25

Postage: 10% (UK), 20% (Europe), 30% (Rest of World). Minimum 30p.

All these publications are available from shops, bookstalls or direct from: Mail Order Department, Platform 5 Publishing Ltd., Wyvern House, Sark Road, SHEFFIELD, S2 4HG, ENGLAND. For a full list of titles available by mail order, please send SAE to the above address.

LIGHT RAIL REVIEW

Over the past 15 years light rail transit has advanced considerably in the UK with the introduction of schemes like Manchester Metrolink and South Yorkshire Supertram, and further important developments are in the pipeline. Light Rail Review takes a comprehensive look at current and future light rail schemes. The editorial content consists of topical articles by recognised authorities in the light rail field, concentrating mainly on UK subjects but also incorporating overseas advances. Much use is made of illustrations, a high proportion of which are in colour making Light Rail Review an informative source of reference which will appeal to both enthusiasts and transport professionals alike. A4 size. Thread sewn.

Light Rail Review 1 . . . . 6.95
Light Rail Review 3 . . . . 7.50
Light Rail Review 4 . . . . 7.50
Light Rail Review 5 **NEW** 7.50

MANX ELECTRIC
by Mike Goodwyn

Manx Electric is the story of the Manx Electric Railway and the Snaefell Mountain Railway. Published to co-incide with the Manx Electric Railway centenary celebrations, it traces the history of the island's electric railway systems in detail and charts the development of the lines and their rolling stock to the present day. As well as a comprehensive history of the two lines, there are detailed chapters on the Rolling Stock, Ticketing, Overhead Equipment, Traffic, Buildings and the Training of Motormen. Author *Mike Goodwyn* is Chairman of the Manx Electric Railway Society and his knowledge of the subject is unsurpassed. He has been connected with the railways for many years, and his achievements include leading the succesful campaign to re-open the section from Laxey to Ramsey in 1977. A4. Thread Sewn. 112 pages including 8 in colour. £8.95.

HIGH SPEED TRAIN TRAILER CARS

HSTs run in formations of 7 or 8 trailer cars with a driving motor brake (power car) at each end. All vehicles are classified mark 3. All trailer cars have BT10 bogies with disc brakes. Heating is by a three-phase supply and vehicles have air conditioning. Max. Speed is 125 mph.

GN4G (TRB)   TRAILER BUFFET FIRST

Dia. GN401. Converted from TRSB by fitting first class seats. Renumbered from 404xx series by subtracting 200. pt. q. 23F.

40204 – 40228. Lot No. 30883 Derby 1976 – 7. 36.12 t.
40231 – 40233. Lot No. 30899 Derby 1978 – 9. 36.12 t.

40204	p	I	IWRR	PM	40212	p	I	ICCT	NL
40205	p	I	IWRR	PM	40213		I	IWRR	PM
40206	p	I	IWRR	PM	40221	p	I	IWRR	PM
40207	p	I	IWRR	PM	40228	p	I	IWRR	PM
40208	p	I	IWRR	PM	40231		I	IWRR	LA
40209	p	I	IWRR	PM	40232		I	ICCT	NL
40210	p	I	IWRR	PM	40233		I	ICCT	NL
40211	p	I	ICCT	NL					

GK2G (TRSB)   TRAILER BUFFET STANDARD

Dia. GK202. pt. Renumbered from 400xx series by adding 400. 35S.

40401 – 40427. Lot No. 30883 Derby 1976 – 7. 36.12 t.
40429 – 40437. Lot No. 30899 Derby 1978 – 9. 36.12 t.

40401	p	I	ICCE	EC	40423	p	I	ICCE	EC
40402	pq	I	ICCE	EC	40424	pq	I	ICCT	LA
40403	pq	I	ICCT	LA	40425	pq	I	ICCT	LA
40414	pq	I	ICCT	NL	40426	pq	I	ICCT	LA
40415	pq	I	ICCT	LA	40427	p	I	ICCE	EC
40416	pq	I	ICCE	EC	40429	pq	I	ICCE	EC
40417	pq	I	ICCT	LA	40430	p	I	ICCE	EC
40418	pq	I	ICCT	LA	40434	pq	I	ICCT	LA
40419	pq	I	ICCE	EC	40435	pq	I	ICCE	EC
40420	pq	I	ICCE	EC	40436	q	I	ICCT	LA
40422	pq	I	ICCE	EC	40437	pq	I	ICCE	EC

GL1G (TRFK)   TRAILER KITCHEN FIRST

Dia. GL101. pt. Reclassified from TRUK. Formerly used in "Pullman" sets, but now used as replacements for out of service TRFBs. pt. 24F.

Lot No. 30884 Derby 1976 – 7. 37 t.

40501	I	IMLX	NL	40511	I	IWRR	LA
40505	I	IECD	NL	40513	I	IMLX	NL

GK1G (TRFM)   TRAILER MODULAR BUFFET FIRST

Dia. GK102. Converted to modular catering from 40719. pt. 17F.

Lot No. 30921 Derby 1978 – 9. 38.16 t.

| 40619 pq | I | ICCT | NL | |

GK1G (TRFB)   TRAILER BUFFET FIRST

Dia. GK101. These vehicles have larger kitchens than the 402xx and 404xx series vehicles, and are used in trains where full meal service is required. They have been renumbered from the 403xx series (in which the seats were unclassified) by adding 400 to previous number. pt. p. q. 17F.

40700 – 40721. Lot No. 30921 Derby 1978 – 9. 38.16 t.
40722 – 40735. Lot No. 30940 Derby 1979 – 80. 38.16 t.
40736 – 40753. Lot No. 30948 Derby 1980 – 1. 38.16 t.
40754 – 40757. Lot No. 30966 Derby 1982. 38.16 t.

40700	I	IMLR	NL	40730	I	IMLR	NL
40701	I	IMLR	NL	40731	I	IWRR	LA
40702	I	IMLR	NL	40732	I	IWRR	LA
40703	I	IWRR	LA	40733	I	IWRR	LA
40704	I	IECD	EC	40734	I	IWRR	LA
40705	I	IECD	EC	40735	I	IECD	EC
40706	I	IECD	EC	40736	I	IWRR	LA
40707	I	IWRR	LA	40737	I	IECD	EC
40708	I	IMLR	NL	40738	I	IWRR	LA
40709	I	IWRR	LA	40739	I	IWRR	PM
40710	I	IWRR	LA	40740	I	IECD	EC
40711	I	IECD	EC	40741	I	IMLR	NL
40712	I	IWRR	LA	40742	I	IWRR	LA
40713	I	IWRR	LA	40743	I	IWRR	LA
40714	I	IWRR	PM	40744	I	IWRR	PM
40715	I	IWRR	PM	40745	I	IWRR	PM
40716	I	IWRR	PM	40746	I	IMLR	NL
40717	I	IWRR	PM	40747	I	IWRR	PM
40718	I	IWRR	LA	40748	I	IECD	EC
40720	I	IECD	EC	40749	I	IMLR	NL
40721	I	IWRR	LA	40750	I	IECD	EC
40722	I	IWRR	LA	40751	I	IMLR	NL
40723	I	IWRR	LA	40752	I	IWRR	PM
40724	I	IWRR	LA	40753	I	IMLR	NL
40725	I	IWRR	LA	40754	I	IMLR	NL
40726	I	IWRR	LA	40755	I	IWRR	LA
40727	I	IWRR	LA	40756	I	IMLR	NL
40728	I	IMLR	NL	40757	I	IWRR	LA
40729	I	IMLR	NL				

GH1G (TF)   TRAILER FIRST

Dia. GH102. pa. 48F 2L.

41003 – 41056. Lot No. 30881 Derby 1976 – 7. 33.66 t.
41057 – 41120. Lot No. 30896 Derby 1977 – 8. 33.66 t.
41121 – 41148. Lot No. 30938 Derby 1979 – 80. 33.66 t.
41149 – 41166. Lot No. 30947 Derby 1980. 33.66 t.
41167 – 41169. Lot No. 30963 Derby 1982. 33.66 t.
41170. Lot No. 30967 Derby 1982. Ex prototype vehicle. 33.66 t.
41178. Lot No. 30882 Derby 1976 – 7. 33.60 t.

Note: 41170 was converted from 41001. 41178 is a prototype refurbished vehicle and has been converted from 42011 which was damaged by fire.

41003	p	I	IWRR	PM	41040	I	IECD	EC
41004		I	IWRR	PM	41041	p	I	IMLR	NL
41005	p	I	IWRR	PM	41042		I	IWRR	PM
41006		I	IWRR	PM	41043		I	IECD	EC
41007	p	I	IWRR	PM	41044		I	IECD	EC
41008		I	IWRR	PM	41045		I	ICCT	LA
41009	p	I	IWRR	PM	41046		I	IMLR	NL
41010		I	IWRR	PM	41049		I	IWRR	PM
41011	p	I	IWRR	PM	41050		I	IWRR	PM
41012		I	IWRR	PM	41051		I	IWRR	LA
41013	p	I	IWRR	PM	41052		I	IWRR	LA
41014		I	IWRR	PM	41055		I	IWRR	LA
41015	p	I	IWRR	PM	41056		I	IWRR	LA
41016		I	IWRR	PM	41057		I	IMLR	NL
41017	p	I	IWRR	PM	41058		I	IMLR	NL
41018		I	IWRR	PM	41059		I	ICCE	EC
41019	p	I	IWRR	PM	41060		I	IWRR	LA
41020		I	IWRR	PM	41061		I	IMLR	NL
41021	p	I	IWRR	PM	41062		I	IMLR	NL
41022		I	IWRR	PM	41063		I	IMLR	NL
41023	p	I	IWRR	LA	41064		I	IMLR	NL
41024		I	IWRR	LA	41065		I	IWRR	LA
41025	p	I	IWRR	LA	41066	p	I	IWRR	LA
41026		I	IWRR	LA	41067		I	IMLR	NL
41027	p	I	IWRR	LA	41068		I	IMLR	NL
41028		I	IWRR	LA	41069		I	IMLR	NL
41029	p	I	IWRR	LA	41070		I	IMLR	NL
41030		I	IWRR	LA	41071		I	IMLR	NL
41031	p	I	IWRR	LA	41072		I	IMLR	NL
41032		I	IWRR	LA	41075		I	IMLR	NL
41033	p	I	IWRR	LA	41076		I	IMLR	NL
41034		I	IWRR	LA	41077		I	IMLR	NL
41035	p	I	IWRR	LA	41078		I	IMLR	NL
41036		I	IWRR	LA	41079		I	IMLR	NL
41037	p	I	IWRR	LA	41080		I	IMLR	NL
41038		I	IWRR	LA	41081		I	ICCE	EC
41039		I	IECD	EC	41082		I	IMLR	NL

41083	I	IMLR	NL	41128	I	IWRR	PM
41084	I	IMLR	NL	41129 p	I	IWRR	PM
41085	I	ICCE	EC	41130	I	IWRR	PM
41086	I	ICCE	EC	41131 p	I	IWRR	LA
41087	I	IECD	EC	41132	I	IWRR	LA
41088	I	IECD	EC	41133 p	I	IWRR	LA
41089	I	IWRR	LA	41134	I	IWRR	LA
41090	I	IECD	EC	41135 p	I	IWRR	LA
41091	I	IECD	EC	41136	I	IWRR	PM
41092	I	IECD	EC	41137 p	I	IWRR	PM
41093	I	IWRR	LA	41138	I	IWRR	PM
41094	I	IWRR	LA	41139 p	I	IWRR	LA
41095	I	ICCE	EC	41140	I	IWRR	LA
41096	I	ICCE	EC	41141 p	I	IWRR	LA
41097	I	IECD	EC	41142	I	IWRR	LA
41098	I	IECD	EC	41143 p	I	IWRR	LA
41099	I	IECD	EC	41144	I	IWRR	LA
41100	I	IECD	EC	41145 p	I	IWRR	PM
41101	I	IWRR	LA	41146	I	IWRR	PM
41102	I	IWRR	LA	41147	I	ICCE	EC
41103	I	IWRR	LA	41148	I	ICCE	EC
41104	I	IWRR	LA	41149	I	ICCE	EC
41105	I	IWRR	PM	41150	I	IECD	EC
41106	I	IWRR	PM	41151	I	IECD	EC
41107	I	ICCE	EC	41152	I	IECD	EC
41108	I	ICCT	LA	41153	I	IMLR	NL
41109	I	ICCT	LA	41154	I	IMLR	NL
41110	I	IWRR	PM	41155	I	IMLR	NL
41111	I	IMLR	NL	41156	I	IMLR	NL
41112	I	IMLR	NL	41157	I	IWRR	LA
41113	I	IMLR	NL	41158	I	IWRR	LA
41114	I	ICCE	EC	41159	I	ICCT	NL
41115	I	IMLR	NL	41160	I	ICCT	NL
41116	I	IWRR	LA	41161	I	ICCT	NL
41117	I	IMLR	NL	41162	I	ICCT	NL
41118	I	IECD	EC	41163	I	ICCT	NL
41119	I	ICCE	EC	41164 p	I	IWRR	LA
41120	I	IECD	EC	41165	I	ICCT	LA
41121 p	I	IWRR	LA	41166	I	ICCT	LA
41122	I	IWRR	LA	41167	I	ICCT	LA
41123 p	I	IWRR	PM	41168	I	ICCT	LA
41124	I	IWRR	PM	41169	I	ICCT	LA
41125	I	IWRR	LA	41170	I	IECD	EC
41126 p	I	IWRR	LA	41178	I	ICCE	EC
41127 p	I	IWRR	PM				

GH2G (TS)   TRAILER STANDARD

Dia. GH203. 76S 2L. pa.

42003 – 42090. Lot No. 30882 Derby 1976 – 7. 33.60 t.
42091 – 42250. Lot No. 30897 Derby 1977 – 9. 33.60 t.

42251 – 42305. Lot No. 30939 Derby 1979 – 80. 33.60 t.
42306 – 42322. Lot No. 30969 Derby 1982. 33.60 t.
42323 – 42341. Lot No. 30983 Derby 1984 – 5. 33.60 t.
42342. Lot No. 30949 Derby 1982. 33.47 t. Converted from TGS.
42343/5. Lot No. 30970 Derby 1982. 33.47 t. Converted from TGS.
42344. Lot No. 30964 Derby 1982. 33.47 t. Converted from TGS.
42346/7/50/1. Lot No. 30881 Derby 1976 – 7. 33.66 t. Converted from TF.
42348/9. Lot No. 30896 Derby 1977 – 8. 33.66 t. Converted from TF.
42353/5 – 7. Lot No. 30967 Derby 1982. Ex prototype vehicles. 33.66 t.
42352/4. Lot No. 30897 Derby 1977. Were TS from 1983 to 1992. 33.66 t.

Notes: 42158 was also numbered 41177 for a time.
42342 – 5 were converted from TGS 44082/95/92/96 respectively.
42346 – 51 were converted from 41053/4, 41073/4, 41047/8 respectively.
42352 – 57 were converted from 41176/1/5/2 – 4 respectively. They were
originally 42142, 42001, 42114, 42000, 42002 and 41002 respectively.

42003	I	IWRR	PM	42040	I	IWRR	LA
42004	I	IWRR	PM	42041	I	IWRR	LA
42005	I	IWRR	PM	42042	I	IWRR	LA
42006	I	IWRR	PM	42043	I	IWRR	LA
42007	I	IWRR	PM	42044	I	IWRR	LA
42008	I	IWRR	PM	42045	I	IWRR	LA
42009	I	IWRR	PM	42046	I	IWRR	LA
42010	I	IWRR	PM	42047	I	IWRR	LA
42012	I	IWRR	PM	42048	I	IWRR	LA
42013	I	IWRR	PM	42049	I	IWRR	LA
42014	I	IWRR	PM	42050	I	IWRR	LA
42015	I	IWRR	PM	42051	I	IWRR	LA
42016	I	IWRR	PM	42052	I	IWRR	LA
42017	I	IWRR	PM	42053	I	IWRR	LA
42018	I	IWRR	PM	42054	I	IWRR	LA
42019	I	IWRR	PM	42055	I	IWRR	LA
42020	I	IWRR	PM	42056	I	IWRR	LA
42021	I	IWRR	PM	42057	I	IECD	EC
42022	I	IWRR	PM	42058	I	IECD	EC
42023	I	IWRR	PM	42059	I	IECD	EC
42024	I	IWRR	PM	42060	I	IWRR	PM
42025	I	IWRR	PM	42061	I	IWRR	PM
42026	I	IWRR	PM	42062	I	IWRR	LA
42027	I	IWRR	PM	42063	I	IECD	EC
42028	I	IWRR	PM	42064	I	IECD	EC
42029	I	IWRR	PM	42065	I	IECD	EC
42030	I	IWRR	PM	42066	I	IWRR	LA
42031	I	IWRR	PM	42067	I	IWRR	LA
42032	I	IWRR	PM	42068	I	IWRR	LA
42033	I	IWRR	LA	42069	I	IWRR	PM
42034	I	IWRR	LA	42070	I	IWRR	PM
42035	I	IWRR	LA	42071	I	IWRR	PM
42036	I	IWRR	LA	42072	I	IWRR	PM
42037	I	IWRR	LA	42073	I	IWRR	PM
42038	I	IWRR	LA	42074	I	IWRR	PM
42039	I	IWRR	LA	42075	I	IWRR	LA

42076	I	IWRR	LA	42129	I	IWRR	LA
42077	I	IWRR	LA	42130	I	ICCT	NL
42078	I	IWRR	LA	42131	I	IMLR	NL
42079	I	IWRR	PM	42132	I	IMLR	NL
42080	I	IWRR	PM	42133	I	IMLR	NL
42081	I	IWRR	LA	42134	I	IWRR	LA
42082	I	IWRR	LA	42135	I	IMLR	NL
42083	I	IWRR	LA	42136	I	IMLR	NL
42084	I	ICCT	NL	42137	I	IMLR	NL
42085	I	ICCT	NL	42138	I	IWRR	PM
42086	I	ICCT	NL	42139	I	IMLR	NL
42087	I	ICCT	NL	42140	I	IMLR	NL
42088	I	ICCT	NL	42141	I	IMLR	NL
42089	I	IWRR	PM	42142	I	IMLR	NL
42090	I	ICCT	NL	42143	I	IWRR	PM
42091	I	ICCT	NL	42144	I	IWRR	PM
42092	I	ICCT	NL	42145	I	IWRR	PM
42093	I	ICCT	NL	42146	I	IECD	EC
42094	I	ICCT	NL	42147	I	IMLR	NL
42095	I	ICCT	NL	42148	I	IMLR	NL
42096	I	IWRR	LA	42149	I	IMLR	NL
42097	I	IWRR	LA	42150	I	IECD	EC
42098	I	IWRR	LA	42151	I	IMLR	NL
42099	I	IWRR	LA	42152	I	IMLR	NL
42100	I	IMLR	NL	42153	I	IMLR	NL
42101	I	IMLR	NL	42154	I	IECD	EC
42102	I	IMLR	NL	42155	I	IMLR	NL
42103	I	ICCE	EC	42156	I	IMLR	NL
42104	I	IECD	EC	42157	I	IMLR	NL
42105	I	ICCT	LA	42158	I	IECD	EC
42106	I	IECD	EC	42159	I	IMLR	NL
42107	I	IWRR	LA	42160	I	IMLR	NL
42108	I	ICCT	LA	42161	I	IMLR	NL
42109	I	ICCT	LA	42162	I	ICCE	EC
42110	I	ICCT	LA	42163	I	IMLR	NL
42111	I	IMLR	NL	42164	I	IMLR	NL
42112	I	IMLR	NL	42165	I	IMLR	NL
42113	I	IMLR	NL	42166	I	ICCE	EC
42115	I	ICCE	EC	42167	I	ICCE	EC
42116	I	ICCE	EC	42168	I	ICCE	EC
42117	I	ICCE	EC	42169	I	ICCE	EC
42118	I	IWRR	PM	42170	I	ICCE	EC
42119	I	IMLR	NL	42171	I	IECD	EC
42120	I	IMLR	NL	42172	I	IECD	EC
42121	I	IMLR	NL	42173	I	ICCE	EC
42122	I	IWRR	LA	42174	I	ICCE	EC
42123	I	IMLR	NL	42175	I	ICCT	LA
42124	I	IMLR	NL	42176	I	ICCT	LA
42125	I	IMLR	NL	42177	I	ICCT	LA
42126	I	IWRR	LA	42178	I	ICCE	EC
42127	I	ICCE	EC	42179	I	IECD	EC
42128	I	ICCE	EC	42180	I	IECD	EC
				42181	I	IECD	EC

42182	I	IECD	EC	42234	I	ICCE	EC
42183	I	IWRR	LA	42235	I	IECD	EC
42184	I	IWRR	LA	42236	I	IWRR	PM
42185	I	IWRR	LA	42237	I	ICCE	EC
42186	I	IECD	EC	42238	I	ICCE	EC
42187	I	ICCE	EC	42239	I	ICCE	EC
42188	I	ICCE	EC	42240	I	IECD	EC
42189	I	ICCE	EC	42241	I	IECD	EC
42190	I	IECD	EC	42242	I	IECD	EC
42191	I	IECD	EC	42243	I	IECD	EC
42192	I	IECD	EC	42244	I	IECD	EC
42193	I	IECD	EC	42245	I	IWRR	LA
42194	I	IMLR	NL	42246	I	ICCE	EC
42195	I	ICCE	EC	42247	I	ICCE	EC
42196	I	IWRR	PM	42248	I	ICCE	EC
42197	I	IWRR	PM	42249	I	ICCE	EC
42198	I	IECD	EC	42250	I	IWRR	LA
42199	I	IECD	EC	42251	I	IWRR	LA
42200	I	IWRR	LA	42252	I	IWRR	LA
42201	I	IWRR	LA	42253	I	IWRR	LA
42202	I	IWRR	LA	42254	I	ICCE	EC
42203	I	IWRR	LA	42255	I	IWRR	PM
42204	I	IWRR	LA	42256	I	IWRR	PM
42205	I	IMLR	NL	42257	I	IWRR	PM
42206	I	IWRR	LA	42258	I	ICCE	EC
42207	I	IWRR	LA	42259	I	IWRR	LA
42208	I	IWRR	LA	42260	I	IWRR	LA
42209	I	IWRR	LA	42261	I	IWRR	LA
42210	I	IMLR	NL	42262	I	ICCE	EC
42211	I	IWRR	PM	42263	I	IWRR	PM
42212	I	IWRR	PM	42264	I	IWRR	LA
42213	I	IWRR	PM	42265	I	IWRR	LA
42214	I	IWRR	PM	42266	I	ICCE	EC
42215	I	IECD	EC	42267	I	IWRR	PM
42216	I	IWRR	LA	42268	I	IWRR	PM
42217	I	ICCE	EC	42269	I	IWRR	PM
42218	I	ICCE	EC	42270	I	ICCE	EC
42219	I	IECD	EC	42271	I	IWRR	LA
42220	I	IMLR	NL	42272	I	IWRR	LA
42221	I	IWRR	LA	42273	I	IWRR	LA
42222	I	ICCT	LA	42274	I	ICCE	EC
42223	I	ICCT	LA	42275	I	IWRR	LA
42224	I	ICCT	LA	42276	I	IWRR	LA
42225	I	IMLR	NL	42277	I	IWRR	LA
42226	I	IECD	EC	42278	I	ICCE	EC
42227	I	IMLR	NL	42279	I	IWRR	LA
42228	I	IMLR	NL	42280	I	IWRR	LA
42229	I	IMLR	NL	42281	I	IWRR	LA
42230	I	IMLR	NL	42282	I	ICCE	EC
42231	I	ICCE	EC	42283	I	IWRR	PM
42232	I	ICCE	EC	42284	I	IWRR	PM
42233	I	ICCE	EC	42285	I	IWRR	PM

42286	I	ICCT	NL	42322	I	ICCT	LA
42287	I	IWRR	LA	42323	I	IECD	EC
42288	I	IWRR	LA	42324	I	IMLR	NL
42289	I	IWRR	LA	42325	I	IWRR	PM
42290	I	ICCT	NL	42326	I	ICCE	EC
42291	I	IWRR	LA	42327	I	IMLR	NL
42292	I	IWRR	LA	42328	I	IMLR	NL
42293	I	IWRR	LA	42329	I	IMLR	NL
42294	I	ICCT	NL	42330	I	ICCE	EC
42295	I	IWRR	LA	42331	I	IMLR	NL
42296	I	IWRR	LA	42332	I	IWRR	PM
42297	I	IWRR	LA	42333	I	IWRR	PM
42298	I	ICCT	NL	42334	I	ICCT	LA
42299	I	IWRR	PM	42335	I	IMLR	NL
42300	I	IWRR	PM	42336	I	ICCE	EC
42301	I	IWRR	PM	42337	I	IMLR	NL
42302	I	ICCT	NL	42338	I	ICCE	EC
42303	I	ICCT	NL	42339	I	IMLR	NL
42304	I	ICCT	NL	42340	I	IECD	EC
42305	I	ICCT	NL	42341	I	IMLR	NL
42306	I	ICCT	LA	42342	I	IWRR	LA
42307	I	ICCT	LA	42343	I	IWRR	LA
42308	I	ICCT	LA	42344	I	IWRR	LA
42309	I	ICCT	LA	42345	I	IWRR	LA
42310	I	ICCT	LA	42346	I	IWRR	PM
42311	I	ICCT	LA	42347	I	IWRR	LA
42312	I	ICCT	LA	42348	I	IWRR	LA
42313	I	ICCT	LA	42349	I	IWRR	LA
42314	I	ICCT	LA	42350	I	IWRR	LA
42315	I	ICCT	LA	42351	I	IWRR	PM
42316	I	ICCT	LA	42352	I	IMLR	NL
42317	I	ICCT	LA	42353	I	ICCE	EC
42318	I	ICCT	LA	42354	I	IECD	EC
42319	I	ICCT	LA	42355	I	IWRR	LA
42320	I	ICCT	LA	42356	I	IWRR	LA
42321	I	ICCT	LA	42357	I	IWRR	LA

GJ2G (TGS)   TRAILER GUARD'S STANDARD

Dia. GJ205. pg. 63S 1L + tip-up seat and wheelchair space.

44000. Lot No. 30953 Derby 1980. 33.47 t.
44001 – 44090. Lot No. 30949 Derby 1980 – 2. 33.47 t.
44091 – 44094. Lot No. 30964 Derby 1982. 33.47 t.
44095 – 44101. Lot No. 30970 Derby 1982. 33.47 t.

§ Fitted with side buffers and drophead buckeye couplings.

44000	I	ICCE	EC	44006	I	IWRR	PM
44001	I	IWRR	LA	44007	I	IWRR	PM
44002	I	IWRR	PM	44008	I	IWRR	PM
44003	I	IWRR	PM	44009	I	IWRR	PM
44004	I	IWRR	PM	44010	I	IWRR	PM
44005	I	IWRR	PM	44011	I	IWRR	LA

44012	I	IWRR	LA	44055	I	ICCE	EC
44013	I	IWRR	LA	44056 §	I	IECD	EC
44014	I	IWRR	LA	44057	I	ICCT	LA
44015	I	IWRR	LA	44058 §	I	IECD	EC
44016	I	IWRR	LA	44059 §	I	IWRR	LA
44017	I	IWRR	LA	44060	I	ICCE	EC
44018	I	IWRR	LA	44061	I	IECD	EC
44019	I	IECD	EC	44062	I	ICCE	EC
44020	I	IWRR	PM	44063	I	IECD	EC
44021 §	I	ICCE	EC	44064	I	IWRR	LA
44022	I	IWRR	LA	44065	I	ICCT	NL
44023	I	IWRR	PM	44066	I	IWRR	LA
44024	I	IWRR	PM	44067	I	IWRR	PM
44025	I	IWRR	LA	44068	I	ICCT	NL
44026	I	IWRR	PM	44069	I	ICCT	NL
44027	I	IMLR	NL	44070	I	IMLR	NL
44028	I	IWRR	LA	44071	I	IMLR	NL
44029	I	IWRR	PM	44072	I	ICCT	NL
44030	I	IWRR	LA	44073	I	IMLR	NL
44031	I	IWRR	LA	44074	I	ICCE	EC
44032	I	IWRR	PM	44075	I	ICCE	EC
44033	I	IWRR	LA	44076	I	ICCT	NL
44034	I	IWRR	LA	44077	I	IECD	EC
44035	I	IWRR	LA	44078	I	ICCE	EC
44036	I	IWRR	PM	44079	I	ICCE	EC
44037	I	IWRR	LA	44080	I	IECD	EC
44038	I	IWRR	LA	44081	I	ICCT	LA
44039	I	IWRR	LA	44083	I	IMLR	NL
44040	I	IWRR	PM	44085	I	IMLR	NL
44041	I	IMLR	NL	44086 §	I	IWRR	LA
44042	I	ICCE	EC	44087	I	ICCT	LA
44043	I	IWRR	LA	44088	I	ICCT	LA
44044	I	IMLR	NL	44089	I	ICCT	LA
44045	I	IECD	EC	44090	I	ICCT	LA
44046	I	IMLR	NL	44091	I	ICCT	LA
44047	I	IMLR	NL	44093	I	IWRR	LA
44048	I	IMLR	NL	44094	I	IECD	EC
44049	I	IWRR	PM	44097 §	I	ICCE	EC
44050	I	IMLR	NL	44098 §	I	IECD	EC
44051	I	IMLR	NL	44099	I	IWRR	LA
44052	I	IMLR	NL	44100	I	ICCE	EC
44053	I	IMLR	NL	44101 §	I	ICCT	LA
44054	I	IMLR	NL				

GH2G (TCSD)   TRAILER CONDUCTOR STANDARD

Dia. GH201. pg. Converted from 44084. Guard's compartment converted to walk-through conductor's compartment with a disabled persons toilet also provided. The car is marshalled adjacent to the buffet.

45084. Lot No. 30949 Derby 1982. 33.47 t.

45084	I	ICCE	EC	

NON-PASSENGER-CARRYING COACHING STOCK

Please note that in this section, that where vehicles have been renumbered more than once, the original number is shown in brackets .

AK51 (RK)   KITCHEN CAR

Dia. AK503. Mark 1. Gas cooking. Converted from RBR. Fluorescent lighting. xd*. ETH 2X.

Note: Kitchen cars have traditionally been numbered in the NPCCS series, but have passenger coach diagram numbers!

Lot No. 30628 Pressed Steel 1960 – 61. 39 t.

80041 (1690)	I	ICHV	BN	

NN   COURIER VEHICLE

Dia. NN504. Converted 1986 – 7 from Mark 1 BSKs. One compartment retained for courier use. Roller shutter doors. xd. ETH 2.

80204 – 6/8/11 – 14/16 – 17/21 – 23/25. Lot No. 30699 Wolverton 1962. Commonwealth bogies. 37 t.
80203/9 – 10/15/20/24. Lot No. 30573 Gloucester 1960. B4 bogies. 35 t.
80218. Lot No. 30427 Wolverton 1958 – 9. B4 bogies. ETH 2. 35 t.
80219. Lot No. 31027 Wolverton 1958. B4 bogies. ETH 2. 35 t.

80204 (35297) x **P**	P	HQ		80217 (35299) x	P	HQ
80206 (35304) x **P**	P	HQ		80218 (35197) x	P	HQ
80207 (35466) x **P**	P	HQ		80219 (35201) x	P	HQ
80208 (35341) x **P**	P	HQ		80220 (35276) x **P**	P	HQ
80211 (35296) x **P**	P	HQ		80221 (35328) x	P	HQ
80212 (35307) x **RX**	P	NC		80222 (35315) x **P**	P	HQ
80213 (35316) x **P**	P	HQ		80223 (35331) x **P**	P	HQ
80214 (35323) x **P**	P	HQ		80224 (35291) x	P	HQ
80216 (35295) x **P**	P	NC		80225 (35327) x **P**	P	HQ

NPX   POST OFFICE GUV

Dia. NP502. Converted 1991 onwards from newspaper vans. Mark 1. Short frames (57'). Originally converted from GUV. Fluorescent lighting, toilets and gangways fitted. Load 14 t. ETH 3X. These vehicles were originally renumbered 85500 – 85534. xd. B5 bogies.

Lot No. 30922 Wolverton or Doncaster 1977 – 8. xe. 31 t (33 t*).

80250 (94008)	**P**	P	EN		80255 (94019)	**P**	P	EN
80251 (94017)	**P**	P	EN		80256 (94013)	**P**	P	EN
80252 (94022)	**P**	P	EN		80257 (94023)	**P**	P	EN
80253 (94018)	**P**	P	EN		80258 (94002)	**P**	P	EN
80254 (94012)	**P**	P	EN		80259 (94005)	**P**	P	EN

NS (POS)   POST OFFICE SORTING VAN

Used in travelling post office (TPO) trains. Mark 1. Various diagrams.

The following lots are vd and have Mark 1 bogies except * – B5 bogies. xd. (subtract 2 t from weight).

80300 – 80305. Lot No. 30486 Wolverton 1959. Dia. NS501. Originally built with nets for collecting mail bags in motion. Equipment now removed. ETH 3X. 36 t.
80306 – 80308. Lot No. 30487 Wolverton 1959. Dia. NS502. ETH 3. 36 t.
80309 – 80314. Lot No. 30661 Wolverton 1961. Dia. NS501. ETH 3. 37 t.
80315 – 80316. Lot No. 30662 Wolverton 1961. Dia. NS501. ETH 3X. 36 t.
80318. Lot No. 30663 Wolverton 1961. Dia. NS501. ETH 3X. 35 t.

80300		**P**	P	HQ	80310	**P**	P	HQ
80301		**P**	P	HQ	80312	**P**	P	HQ
80303	*	**P**	P	DY	80313	**P**	P	HQ
80305		**P**	P	HQ	80314 *	**P**	P	NC
80306		**P**	P	NC	80315	**P**	P	HQ
80308	*	**P**	P	HT	80316 *	**P**	P	NC
80309	*	**P**	P	HT	80318 *	**P**	P	NC

The following lots are pressure ventilated and have B5 bogies.

80319 – 80327. Dia. NS504. Lot No. 30778 York 1968 – 9. xd. ETH 4. 35 t.
80328 – 80338. Dia. NS505. Lot No. 30779 York 1968 – 9. xd. ETH 4. 35 t.
80339 – 80355. Dia. NS506. Lot No. 30780 York 1968 – 9. xd. ETH 4. 35 t.

80319	**P**	P	CA	80338	**P**	P	HT
80320	**P**	P	CA	80339	**P**	P	HT
80321	**P**	P	CA	80340	**P**	P	DY
80322	**P**	P	CA	80341	**P**	P	EN
80323	**P**	P	CA	80342	**P**	P	EN
80324	**P**	P	CA	80343	**P**	P	DY
80325	**P**	P	DY	80344	**P**	P	CA
80326	**P**	P	DY	80345	**P**	P	HT
80327	**P**	P	DY	80346	**P**	P	HT
80328	**P**	P	EN	80347	**P**	P	CA
80329	**P**	P	DY	80348	**P**	P	HT
80330	**P**	P	CA	80349	**P**	P	DY
80331	**P**	P	HT	80350	**P**	P	HT
80332	**P**	P	CA	80351	**P**	P	DY
80333	**P**	P	HT	80352	**P**	P	HT
80334	**P**	P	DY	80353	**P**	P	EN
80335	**P**	P	HT	80354	**P**	P	EN
80336	**P**	P	HT	80355	**P**	P	DY
80337	**P**	P	HT				

Name: 80320 The Borders Mail

80356 – 80380. Lot No. 30839 York 1972 – 3. Dia. NS501. Pressure ventilated. Fluorescent lighting. B5 bogies. xd. ETH 4X. 37 t.

80356	**P**	P	EN	80357	**P**	P	EN

80358	**P**	**P**	HT	80370	**P**	**P**	EN
80359	**P**	**P**	HT	80371	**P**	**P**	EN
80360	**P**	**P**	HT	80372	**P**	**P**	EN
80361	**P**	**P**	HT	80373	**P**	**P**	EN
80362	**P**	**P**	HT	80374	**P**	**P**	DY
80363	**P**	**P**	HT	80375	**P**	**P**	DY
80364	**P**	**P**	HT	80376	**P**	**P**	EN
80365	**P**	**P**	HT	80377	**P**	**P**	EN
80366	**P**	**P**	EN	80378	**P**	**P**	EN
80367	**P**	**P**	EN	80379	**P**	**P**	EN
80368	**P**	**P**	EN	80380	**P**	**P**	DY
80369	**P**	**P**	EN				

Name: 80367 M.G. Berry

80381 – 80395. Lot No. 30900 Wolverton 1977. Dia NS531. Converted from SK. Pressure ventilated. Fluorescent lighting. B5 bogies. xd. ETH 4X. 38 t.

80381	(25112)	**P**	**P**	EN	80389	(25103)	**P**	**P**	EN
80382	(25109)	**P**	**P**	EN	80390	(25047)	**P**	**P**	EN
80383	(25033)	**P**	**P**	EN	80391	(25089)	**P**	**P**	EN
80384	(25078)	**P**	**P**	EN	80392	(25082)	**P**	**P**	EN
80385	(25083)	**P**	**P**	EN	80393	(25118)	**P**	**P**	EN
80386	(25099)	**P**	**P**	EN	80394	(25156)	**P**	**P**	EN
80387	(25045)	**P**	**P**	EN	80395	(25056)	**P**	**P**	DY
80388	(25088)	**P**	**P**	EN					

NT (POT)   POST OFFICE STOWAGE VAN

Mark 1. Open vans used for stowage of mail bags in conjunction with POS. Various diagrams.

Lot No. 30488 Wolverton 1959. Dia. NT502. Originally built with nets for collecting mail bags in motion. Equipment now removed. B5 bogies. xd. ETH 3. 35 t.

80400	**P**	**P**	EN	80402	**P**	**P**	EN
80401	**P**	**P**	EN				

The following twelve vehicles were converted at York from BSK to lot 30143 (80403) and 30229 (80404 – 80414). No new lot number was issued. Dia. NT503. B5 bogies. xd. 35 t. (*dia. NT501 BR2 bogies 38 t. ETH 3 (3X*).

80403	(34361)		**P**	**P**	DY	80411	(35003)	* **P**	**P**	CA
80404	(35014)		**P**	**P**	DY	80412	(35002)	* **P**	**P**	CA
80405	(35009)		**P**	**P**	DY	80413	(35004)	* **P**	**P**	CA
80406	(35022)		**P**	**P**	DY	80414	(35005)	* **P**	**P**	DY

Lot No. 30781 York 1968. Dia. NT505. Pressure ventilated. B5 bogies. xd. ETH 4. 34 t.

80415	**P**	**P**	EN	80420	**P**	**P**	HT
80416	**P**	**P**	EN	80421	**P**	**P**	HT
80417	**P**	**P**	EN	80422	**P**	**P**	HT
80418	**P**	**P**	EN	80423	**P**	**P**	EN
80419	**P**	**P**	EN	80424	**P**	**P**	EN

Lot No. 30840 York 1973. Dia. NT504. Pressure ventilated. fluorescent lighting.
B5 bogies. xd. ETH 4X. ·35 t.

80425	P	P	EN		80428	P	P	EN
80426	P	P	HT		80429	P	P	EN
80427	P	P	HT		80430	P	P	EN

Lot No. 30901 Wolverton 1977. converted from SK. Dia. NT521. Pressure ventilated. Fluorescent lighting. B5 bogies. xd. ETH 4X. 35 t.

80431 (25104)	P	P	DY		80436 (25077)	P	P	EN
80432 (25071)	P	P	HT		80437 (25068)	P	P	EN
80433 (25150)	P	P	HT		80438 (25139)	P	P	DY
80434 (25119)	P	P	HT		80439 (25127)	P	P	DY
80435 (25117)	P	P	EN					

NU (BPOT)   BRAKE POST OFFICE STOWAGE VAN

As NT but with brake. Mark 1.

Lot No. 30782 York 1968. Dia. NU502. Pressure ventilated. B5 bogies. xd.
ETH 4. 36 t.

80456	P	P	EN		80458	P	P	EN
80457	P	P	EN					

ND (BG)   GANGWAYED BRAKE VAN (90 mph)

Dia. ND501. These vans are built on short frames (57'). Load 10t. BR1 bogies.
d. ETH 1 (1X*). Only fifteen vans remain in this once numerous series, but the
full lot number list is listed here for reference purposes with renumbered vehicles.

b − (Dia. NB501). High security letter mail van. Converted at WB from BG 1985.
Gangways removed. vd. 84382/7/461/77 formerly renumbered 80460 − 3, but
since given original number to prevent identification by potential train robbers!

80501 − 80529. Lot No. 30009 Derby 1952 − 3. 31 t.
80537 − 80565. Lot No. 30039 Derby 1954. 31 t.
80570 − 80596. Lot No. 30040 Wolverton 1954 − 5. 32 t.
80597 − 80671. Lot No. 30046 York 1954. 31.5 t.
80672 − 80724. Lot No. 30136 Metro-Cammell 1955. 31.5 t.
80726 − 80802. Lot No. 30140 BRCW 1955 − 6. 31.5 t.
80803 − 80852. Lot No. 30144 Cravens 1955. 31.5 t.
80855 − 80962. Lot No. 30162 Pressed Steel 1956 − 7. 32 t.
80965 − 80999. Lot No. 30173 York 1956. 31.5 t.

80730		NXXZHQ			80926 x		P	HQ
80889 x*	P	P	HQ				

NZ (DLV)   DRIVING BRAKE VAN (110 mph)

Dia. NZ501. Mark 3B. Air conditioned. T4 bogies. ae. dg. Cab to shore communication. ETH 5X.

Lot No. 31042 Derby 1988. 45.18 t.

82101	I	IWCX	MA	82127	I	IWCX	MA
82102	I	IWCX	OY	82128	I	IWCX	OY
82103	I	IWCX	OY	82129	I	IWCX	OY
82104	I	IWCX	PC	82130	I	IWCX	WB
82105	I	IWCX	MA	82131	I	IWCX	WB
82106	I	IWCX	OY	82132	I	IWCX	OY
82107	I	IWCX	PC	82133	I	IWCX	OY
82108	I	IWCX	MA	82134	I	IWCX	WB
82109	I	IWCX	MA	82135	I	IWCX	MA
82110	I	IWCX	WB	82136	I	IWCX	MA
82111	I	IWCX	PC	82137	I	IWCX	WB
82112	I	IWCX	WB	82138	I	IWCX	PC
82113	I	IWCX	OY	82139	I	IWCX	WB
82114	I	IWCX	WB	82140	I	IWCX	WB
82115	I	IWCX	WB	82141	I	IWCX	MA
82116	I	IWCX	MA	82142	I	IWCX	WB
82117	I	IWCX	PC	82143	I	IWCX	OY
82118	I	IWCX	PC	82144	I	IWCX	OY
82119	I	IWCX	WB	82145	I	IWCX	OY
82120	I	IWCX	WB	82146	I	IWCX	MA
82121	I	IWCX	MA	82147	I	IWCX	MA
82122	I	IWCX	OY	82148	I	IWCX	OY
82123	I	IWCX	OY	82149	I	IWCX	WB
82124	I	IWCX	PC	82150	I	IWCX	WB
82125	I	IWCX	OY	82151	I	IWCX	OY
82126	I	IWCX	MA	82152	I	IWCX	OY

NZ (DLV)   DRIVING BRAKE VAN (140 mph)

Dia. NZ502. Mark 4. Air conditioned. Swiss-built (SIG) bogies. ae. dg. Cab to shore communication. ETH 6X.

Lot No. 31043 Metro-Cammell 1988. 45.18 t.

82200	I	IECX	BN	82216	I	IECX	BN
82201	I	IECX	BN	82217	I	IECX	BN
82202	I	IECX	BN	82218	I	IECX	BN
82203	I	IECX	BN	82219	I	IECX	BN
82204	I	IECX	BN	82220	I	IECX	BN
82205	I	IECX	BN	82221	I	IECX	BN
82206	I	IECX	BN	82222	I	IECX	BN
82207	I	IECX	BN	82223	I	IECX	BN
82208	I	IECX	BN	82224	I	IECX	BN
82209	I	IECX	BN	82225	I	IECX	BN
82210	I	IECX	BN	82226	I	IECX	BN
82211	I	IECX	BN	82227	I	IECX	BN
82212	I	IECX	BN	82228	I	IECX	BN
82213	I	IECX	BN	82229	I	IECX	BN
82214	I	IECX	BN	82230	I	IECX	BN
82215	I	IECX	BN	82231	I	IECX	BN

NZ (DLV)   DRIVING BRAKE VAN (Exptl.)

Dia. NZ503. This is in effect locomotive 33115 with its traction equipment isolated which has been fitted with TGV bogies with third-rail pick-up shoes. It works coupled to a Class 73 locomotive which has its shoegear removed.

| 83301 | (33115) | I | GPSG | SL | |

NB/NC/ND (BG) GANGWAYED BRAKE VAN (90 mph)

Class continued from 80994. Note: All 84XXX vehicles were renumbered from the 81XXX series by adding 3000 to the original number.

84000 – 84014. Lot No. 30173 York 1956. 31.5 t.
84015 – 84053. Lot No. 30224 Cravens 1956. 31.5 t.
84055 – 84179. Lot No. 30228 Metro-Cammell 1957 – 8. 31.5 t.
84180 – 84204. Lot No. 30234 Cravens 1956 – 7. 31.5 t.
84205 – 84265. Lot No. 30163 Pressed Steel 1957. 31.5 t.
84266 – 84312. Lot No. 30323 Pressed Steel 1957. 32 t.
84313 – 84497. Lot No. 30400 Pressed Steel 1957 – 8. 32 t.
84498 – 84572. Lot No. 30484 Pressed Steel 1958. 32 t.
84573 – 84592. Lot No. 30715 Gloucester 1962. 31 t.
84594 – 84612. Lot No. 30716 Gloucester 1962. 31 t.
84613 – 84628. Lot No. 30725 Gloucester 1962 – 3. 31 t.

84044	x		P	HQ		84445	x		P	HQ
84197			P	HQ		84456	x		P	HQ
84234	x*	**P**	P	HQ		84477	b		P	EN
84382	b		P	EN		84499	x		P	HQ
84387	b		P	EN		84510	x	**P**	P	HQ
84399	x		P	HQ		84519			P	HQ
84419	x		P	HQ		84531	x*		P	HQ
84428			P	HQ						

NE/NH (BG) 100/110 mph GANGWAYED BRAKE VAN

As ND but rebogied with B4 bogies suitable for 100 mph – NE (110 mph with special maintenance – NH). d. ETH 1 (1X* and NHA). For lot numbers refer to original number series. Deduct 1.5t from weights. All NHA are a*pg.

92067	(81243) a to	I	P		HQ
92100	(81391) a to	I	ICHX		BN
92103	(81400) x to		P		BK
92105	(81405) a to	**P**	P		HT
92110	(81426) x*to	**RX**	P		EN
92111	(81432) NHA	I	ICCX		PC
92112	(81440) x to	**P**	P		DY
92113	(81442) a to		P		CA
92114	(81443) NHA	I	ICCX		PC
92116	(81450) x to	I	ICHX		BN
92119	(81454) x to	**P**	P		BK
92120	(81455) x to		P		CA
92121	(81457) x*to	I	P		BK

92122 (81459) x*to	**P**	P	CA
92124 (81465) x	**P**	P	BK
92125 (81470) a to	**I**	IXXZ	DY
92126 (81471) a pg	**I**	IXXZ	DY
92127 (81473) x*pg	**RX**	P	CA
92130 (81480) a to		P	HQ
92131 (81481) a to	**P**	P	HT
92132 (81482) a to	**P**	P	BK
92133 (81484) a to		P	HQ
92134 (81485) a to		P	BK
92135 (81486) a to		P	BK
92136 (81487) a to	**P**	P	CA
92137 (81488) a to	**P**	P	HT
92138 (81490) a to	**P**	P	BK
92139 (81491) a to	**P**	P	HT
92140 (81492) a to		P	BK
92144 (81496) a to	**P**	P	HT
92146 (81498) NHA	**I**	IWCX	WB
92147 (81500) a to	**I**	P	BK
92148 (81506) a to	**P**	IXXZ	DY
92152 (81518) a to		P	CA
92154 (81522) x to	**P**	P	HT
92155 (81525) a*pg	**I**	ICCX	DY
92156 (81529) x*to	**I**	P	CA
92157 (81532) x*pg	**RX**	P	CA
92158 (81533) a to		P	HQ
92159 (81534) NHA	**I**	IWCX	WB
92160 (81537) x to	**P**	P	HQ
92161 (81538) a to	**P**	P	HQ
92162 (81539) a to	**P**	P	HQ
92166 (81550) a to		P	HQ
92168 (81555) a to	**P**	P	HQ
92169 (81556) x	**P**	P	CA
92170 (81559) x to		P	CA
92172 (81562) a pg	**I**	P	HQ
92173 (81565) x to	**P**	P	CA
92174 (81567) NHA	**I**	IWCX	PC
92175 (81568) a pg	**I**	IXXZ	LA
92176 (81569) x to	**P**	P	CA
92177 (81572) a to	**P**	P	HQ
92178 (81574) x		P	HQ
92179 (81580) x to	**P**	P	CA
92181 (81582) x to		P	CA
92182 (81585) x to	**P**	P	CA
92183 (81588) a*pg	**I**	IXXZ	DY
92184 (81589) a to	**P**	P	HQ
92186 (81595) a to	**I**	P	HQ
92187 (81597) a to	**P**	P	HQ
92188 (81598) a to		IXXZ	LA
92190 (81600) a to	**I**	IXXZ	DY
92192 (81602) x to		P	CA
92193 (81604) a pg	**I**	P	HQ

92194 (81606) a to	I	IWRX	LA
92195 (81607) a to	I	IXXZ	DY
92196 (81609) a to	I	IXXZ	DY
92197 (81610) a to	I	IWRX	LA
92198 (81613) a to		P	HQ
92200 (81615) x to	P	P	BK
92203 (81621) a to	I	P	HQ
92205 (81623) a to	P	P	HQ
92206 (81624) x		P	CA
92207 (81627) x to	P	P	CA
92208 (84403) x		P	BK
92211 (81267) x	P	P	EN
92212 (80937) x	P	P	EN
92215 (80911) x		P	CA
92224 (84277) x	P	P	DY
92226 (80879) x	P	P	EN
92227 (84337) x		P	CA
92228 (80921) x		P	CA
92229 (80902) x	P	P	CA
92230 (81423) x	P	P	CA
92232 (80870) x		P	BK
92233 (80890) x	P	P	CA
92234 (84336) x	RX	P	BK
92235 (80908) x		P	BK
92236 (80909) x	RX	P	BK
92238 (84563) x	P	P	BK
92239 (81170) x	P	P	CA
92242 (80857) x	RX	P	CA
92243 (84489) x	P	P	EN
92244 (84248) x	P	P	DY
92248 (80935) x	P	P	DY
92249 (84511) x	P	P	EN
92251 (84425) x		P	CA
92252 (80959) x	P	P	EN
92257 (80955) x	P	P	BK
92258 (84346) x	P	P	EN
92259 (84313) x	P	P	DY
92260 (84104) x	P	P	EN
92261 (80988) x*	P	P	DY
92262 (84240) x*	P	P	EN
92263 (84325) x		P	BK
92264 (84239) x*	P	P	CA
92265 (80945) x	P	P	BK
92266 (84515) x	P	P	BK
92267 (84404) x		P	CA
92268 (84352) x		P	DY
92269 (84010) x		P	CA
92270 (84289) x	P	P	DY
92271 (80962) x*	P	P	BK
92272 (84262) x*	P	P	EN

NE (BG)   100 mph GANGWAYED BRAKE VAN

As ND but rebogied with Commonwealth bogies suitable for 100 mph. xe. ETH 1
(1X*). For lot numbers refer to original number series. Add 1.5 t to weights to
allow for the increased weight of the Commonwealth bogies.

Number			Depot
92301 (80737) x		P	CA
92302 (84501) x	RX	P	CA
92303 (84427) x		P	BK
92306 (84217) x*	P	P	BK
92307 (80805) *		P	BK
92309 (84043) x*	RX	P	EN
92311 (84453) x	P	P	DY
92312 (84548) x	RX	P	BK
92313 (80992) x*		P	CA
92314 (80777) x*	P	P	EN
92316 (80980) x*	P	P	BK
92317 (80836) x*	RX	P	CA
92318 (80847) x*		P	CA
92319 (84055) x	P	P	HT
92321 (84566) x	P	P	BK
92322 (80771) x*	RX	P	CA
92323 (80832) x*	P	P	EN
92324 (84087) x	P	P	EN
92325 (80791) x	P	P	BK
92328 (80999) x*	P	P	BK
92329 (84001) x*	P	P	BK
92330 (80995) x*	P	P	BK
92332 (80845) x*	RX	P	DY
92333 (80982) x*	P	P	EN
92334 (80983) x*	P	P	CA
92335 (80973) x*		P	CA
92337 (84140) x*	RX	P	BK
92339 (84530) x	P	P	BK
92340 (84059) x*	P	P	EN
92341 (84316) x	P	P	EN
92343 (84505) x	P	P	BK
92344 (84154) x*	P	P	BK
92345 (84083) x*	P	P	EN
92346 (84091) x	P	P	EN
92347 (84326) x	RX	P	EN
92348 (84075) x*	P	P	EN
92349 (84178) x	P	P	EN
92350 (84049) x*	P	P	HT
92351 (84174) x	RX	P	EN
92353 (84323) x	P	P	CA
92355 (84517) x	RX	P	DY
92356 (84535) x		P	BK
92357 (84136) x	RX	P	EN
92358 (84393) x	RX	P	BK
92362 (84188) x	P	P	HT

92363 (84294) x	P	P	EN
92364 (84030) x*	P	P	BK
92365 (84122) x	RX	P	EN
92366 (84551) x	RX	P	CA
92369 (80960) x*		P	CA
92370 (84324) x	RX	P	EN
92371 (80856) x*		P	BK
92374 (84317) x		P	CA
92377 (80928) x*	RX	P	CA
92379 (80914) x*	RX	P	CA
92380 (84247) x*	P	P	EN
92381 (84476) x	RX	P	EN
92382 (84561) x	RX	P	EN
92384 (80893) x	P	P	EN
92385 (84261) x*	P	P	BK
92387 (84380) x		P	BK
92388 (80868) x*	P	P	BK
92389 (84026) x	P	P	EN
92390 (80834) x*	RX	P	CA
92391 (80790) x	P	P	DY
92392 (80861) x*	P	P	HT
92395 (84274) x		P	BK
92396 (84430) x		P	HQ
92398 (80859) x*	P	P	EN
92399 (80781) x*	P	P	DY
92400 (84211) x*		P	CA
92401 (84280) x	RX	P	CA
92402 (84099) x*	P	P	EN
92403 (84273) x	P	P	EN
92404 (84051) x*		P	BK
92405 (84320) x	P	P	HQ
92406 (84475) x	P	P	EN
92407 (84363) x		P	CA
92408 (84351) x		P	BK
92409 (84370) x*		P	CA
92410 (84469) x		P	BK
92411 (84252) x*	P	P	BK
92412 (84354) x	P	P	CA
92413 (84472) x	P	P	CA
92414 (84458) x		P	CA
92415 (84388) x	RX	P	EN
92416 (84250) x*	P	P	CA
92417 (80885) x*	RX	P	CA
92418 (84512) x	RX	P	BK

NF (BG)   100/110 mph GANGWAYED BRAKE VAN

Vehicles with emergency equipment removed. For details and lot numbers refer to original number series. For last number add 400 to 925xx vehicles, but subtract 500 for other vehicles. (Does not apply to 92530).

b – NBV. See earlier.

92503 (80864) x*to	**P**	P	EN	
92505 (80876) x*to	**I**	P	EN	
92509 (80897) x*to	**RX**	P	BK	
92510 (80900) x*to	**I**	P	DY	
92513 (80916) x*pg	**RX**	P	BK	
92518 (90941) x*to	**I**	P	BK	
92521 (80956) x*to	**I**	P	BK	
92530 (84461) b	**RX**	P	EN	
92542 (81207) a to	**P**	P	CA	
92547 (81216) a to	**P**	P	DY	
92550 (81220) a pg	**RX**	P	CA	
92553 (81223) a to	**P**	P	CA	
92568 (81244) a to	**RX**	P	CA	
92577 (81258) a to		P	BK	
92584 (81268) a to	**P**	P	EN	
92602 (81394) a to		P	HQ	
92604 (81401) a to	**P**	P	DY	
92606 (81409) x to	**P**	P	CA	
92607 (81410) x to	**RX**	P	CA	
92608 (81411) a to	**P**	P	CA	
92609 (81413) a to	**RX**	P	CA	
92615 (81444) a to		P	BK	
92617 (81451) x	**P**	P	CA	
92618 (81452) a to		P	EN	
92629 (81479) a to	**P**	P	CA	
92641 (81493) a to	**P**	P	CA	
92642 (81494) a to		P	BK	
92643 (81495) a to	**P**	P	CA	
92645 (81497) a to	**P**	P	EN	
92649 (81509) x		P	BK	
92650 (81514) x to	**P**	P	BK	
92651 (81516) a to	**I**	P	CA	
92663 (81540) a to		P	HQ	
92664 (81541) a to	**RX**	P	CA	
92665 (81546) a to		P	EN	
92704 (81622) a to	**P**	P	HQ	
92709 (80873) x	**RX**	P	BK	
92714 (84504) x	**RX**	P	BK	
92716 (81376) x	**P**	P	CA	
92717 (80877) x	**P**	P	CA	
92718 (84314) x	**P**	P	EN	
92720 (80924) x	**P**	P	CA	
92721 (80888) x	**P**	P	DY	
92722 (80887) x	**P**	P	CA	
92723 (80932) x	**RX**	P	BK	
92725 (80891) x	**RX**	P	EN	
92740 (80703) x	**P**	P	EN	
92741 (80943) x		P	BK	
92746 (80929) x	**P**	P	EN	
92747 (84536) x	**P**	P	EN	
92750 (81235) x	**RX**	P	BK	
92753 (80936) x	**P**	P	BK	

▲ **Mark 3B Stock.** Open first No. 11099 at Carlisle on 15th August 1993.
Kevin Conkey

▼ **Mark 4 Stock.** Open standard (end) (TSOE) No. 12223 in the formation of a northbound train at Durham on 17th April 1993.　*John Augustson*

▲ **Non-Passenger-Carrying Coaching Stock.** Parcels-liveried courier van No. 80214 at Newcastle on 23rd July 1991. *John Augustson*

▼ Post Office general utility van (GUV) No. 80252 at Plymouth on 27th July 1993. *Stephen Widdowson*

Post Office sorting van No. 80320 'The Borders Mail' stabled at Carlisle on 26th June 1992. *Dave McAlone*

Mark 3 DVT No. 82132 leads the 15.30 Glasgow Central – Euston at Penrith on 13th August 1993.
Kevin Conkey

Mark 4 DVT No. 82211 approaches Darlington at the head of a Newcastle – Kings Cross service on 14th April 1993.

John Augustson

▲ Commonwealth bogied brake van No. 92381 in Rail Express Systems livery at York on 11th January 1992. *Dave McAlone*

▼ B4 bogied brake van No. 92521 at Carlisle on 3rd May 1993. *Dave McAlone*

Propelling control vehicle (PCV) No. 94300 stands outside the Technical centre at Derby shortly after being converted from a Class 307 EMU driving trailer, on 20th August 1993.
Brian Morrison

▲ General Utility van No. 95144 at Bolton on 25th June 1993.
Vincent Eastwood

▼ Former Newspaper van No. 95303 at Reading on 11th June 1993. This vehicle carries plain blue livery.
C J Marsden

92754	(80894) x	**P**	P	BK
92755	(80871) x	**P**	P	EN
92800	(81200) x	**RX**	P	CA
92804	(84339) x	**RX**	P	EN
92805	(81590) x	**RX**	P	CA
92808	(80784) x*	**RX**	P	EN
92810	(81105) x	**P**	P	EN
92815	(80848) x*	**P**	P	BK
92820	(84166) x		P	BK
92826	(84270) x		P	BK
92827	(80842) x*		P	HT
92831	(84365) x	**RX**	P	BK
92842	(84397) x	**P**	P	BK
92852	(81182) x	**RX**	P	CA
92854	(84353) x	**RX**	P	BK
92859	(81275) x	**RX**	P	DY
92860	(84431) x	**RX**	P	CA
92861	(81463) x	**P**	P	BK
92867	(81293) x	**RX**	P	BK
92868	(84334) x	**RX**	P	CA
92872	(84362) x	**P**	P	CA
92873	(84528) x	**RX**	P	BK
92875	(84335) x		P	HQ
92876	(81374) x	**RX**	P	EN
92883	(84429) x	**RX**	P	BK
92886	(80843) x*	**RX**	P	CA
92893	(80701) x	**P**	P	EN
92894	(81322) x	**RX**	P	CA
92897	(80700) x*	**P**	P	BK

NE/NH (BG)   100/110 mph GANGWAYED BRAKE VAN

Renumbered from 92xxx series by adding 900 to number to avoid conflict with
Class 92 locos.

92901	(80855) NHA	**I**	IWCX	WB
92902	(80858) x*to	**I**	P	BK
92904	(80867) x*to	**I**	P	EN
92906	(80878) NHA	**I**	ICCX	PC
92907	(80880) x*pg	**RX**	P	CA
92908	(80895) NHA	**I**	IWCX	WB
92912	(80910) a*pg	**I**	ICCX	MA
92914	(80923) x*pg	**I**	P	BK
92915	(80927) x*to	**P**	P	BK
92916	(80930) x*pg	**P**	P	CA
92917	(80940) x*to	**RX**	P	BK
92919	(80944) x*pg	**I**	P	BK
92920	(80950) x*pg	**I**	P	BK
92922	(80958) x*pg	**RX**	P	EN
92923	(80971) a*pg	**I**	IWCX	WB
92926	(81060) NHA	**I**	ICCX	PC

92927 (81061) NHA	I	IWCX	WB
92928 (81064) NHA	I	ICCX	PC
92929 (81077) NHA	I	ICCX	PC
92931 (81102) NHA	I	IWCX	PC
92932 (81117) NHA	I	IWCX	WB
92933 (81123) NHA	I	IWCX	MA
92934 (81142) NHA	I	IWCX	WB
92935 (81150) a*pg	I	IWCX	WB
92936 (81158) NHA	I	IWCX	WB
92937 (81165) NHA	I	IWCX	WB
92938 (81173) NHA	I	IWCX	WB
92939 (81175) NHA	I	IWCX	WB
92940 (81186) a pg	I	IWRX	LA
92941 (81205) a to		P	CA
92943 (81208) a to	P	P	HT
92944 (81209) a to	I	P	BK
92945 (81210) a to		P	CA
92946 (81214) NHA	I	IWCX	WB
92948 (81218) NHA	I	IWCX	WB
92954 (81224) a to	P	P	HT
92955 (81225) a to	RX	P	CA
92956 (81226) a to	P	P	CA
92957 (81227) a to	I	ICHX	BN
92958 (81228) a to	RR	P	BK
92959 (81229) a to	P	P	DY
92960 (81230) a to	P	P	CA
92961 (81231) a	I	IANX	NC
92962 (81232) a to		P	BK
92964 (81236) a to	P	P	CA
92965 (81237) a		P	BK
92966 (81238) a to		P	BK
92969 (81245) a to	RX	P	CA
92971 (81249) a to		P	HQ
92972 (81253) a to		P	EN
92973 (81254) a to	P	P	CA
92974 (81255) a to	P	P	CA
92975 (81256) a to	P	P	HT
92976 (81257) a to	RX	P	BK
92978 (81259) a to	RX	P	EN
92979 (81260) a to	P	P	CA
92981 (81264) a to	P	P	CA
92982 (81265) a to	P	P	HT
92983 (81266) a to	RX	P	HT
92986 (81282) a to	I	IANX	NC
92987 (81283) x*to	I	P	BK
92988 (81284) a to	I	ICCX	PC
92989 (81303) a to	I	IXXZ	HQ
92990 (81305) a to		P	HQ
92991 (81308) a to	I	IANX	NC
92992 (81309) a to	I	P	CA
92994 (81367) a to	I	IXXZ	DY
92995 (81375) a to	P	P	CA

92996 (81377) a to		P	CA
92997 (81378) a to		P	BK
92998 (81381) a to	I	IWCX	WB
92999 (81383) x*pg	I	P	BK

NJ/NK/NX (GUV)   GENERAL UTILITY VAN

Mark 1. Short frames. Load 14 t. All vehicles are through steam piped only (Dia. NK501) unless otherwise stated. Screw couplings. Electric wired vehicles or steam piped and electric wired vehicles are Dia. NJ501. All vehicles have BR Mark 2 bogies. ETH 0 or 0X.

93078 – 93499. Lot No. 30417 Pressed Steel 1958 – 9. 30 t.
93501 – 93519. Lot No. 30343 York 1957. 30 t.
93521 – 93654. Lot No. 30403 York/Glasgow 1958 – 60. 30 t.
93655 – 93834. Lot No. 30565 Pressed Steel 1959. 30 t.
93835 – 93984. Lot No. 30616 Pressed Steel 1959 – 60. 30 t.

93131	vr*	**P**	P	HQ	93717	vy	**B**	P	HQ
93135	vr*	**B**	P	HQ	93720	vy	**P**	P	HQ
93267	vr*	**P**	P	HQ	93823	vr	**B**	P	HQ
93273	xr*	**B**	P	HQ	93830	vr	**B**	P	HQ
93394	vr*	**P**	P	HQ	93847	vr*	**P**	P	HQ
93446	vr	**B**	P	HQ	93849	vr*	**B**	P	HQ
93474	vr*	**P**	P	CA	93859	vr*	**B**	P	HQ
93511	vr*	**P**	P	HQ	93881	vr*	**B**	P	HQ
93515	vr*	**B**	P	HQ	93886	vr*	**B**	P	HQ
93525	vr*	**P**	P	HQ	93905	vr*	**P**	P	HQ
93541	vr*	**B**	P	HQ	93935	vr*	**P**	P	HQ
93562	vr*	**B**	P	HQ	93952	vr*	**B**	P	HQ
93585	vr*	**P**	P	HQ	93955		**B**	P	HQ
93660	vr*	**P**	P	HQ	93962	vr*	**B**	P	HQ
93701	vr*	**B**	P	HQ	93973	vr*	**B**	P	HQ
93711	vy	**B**	P	CA	93979	vr*	**B**	P	HQ
93714	vy	**B**	P	HQ					

NLX   NEWSPAPER VAN

Dia. NL501. Mark 1. Short frames (57'). Converted from GUV. Fluorescent lighting, toilets and gangways fitted. Load 14 t. ETH 3X. These vehicles were originally renumbered 85500 – 85534. Not now used for News traffic. B5 bogies.

Lot No. 30922 Wolverton or Doncaster 1977 – 8. xe. 31 t (33 t*).

94003 (93999)	**RX**	P	CA	94020 (86220)	**P**	P	CA
94004 (86156)	**P**	P	CA	94021 (86204)	**B**	P	BK
94006 (86202)	**B**	P	BK	94024 (86106)	**B**	P	BK
94007 (86572)	**B**	P	BK	94025 (86377)	**P**	P	BK
94009 (86144)	**P**	P	CA	94026 (86703)	**P**	P	EN
94010 (86151)	**RX**	P	CA	94027 (86732)	**P**	P	EN
94011 (86437)	**B**	P	EN	94028 (86733)	**RX**	P	EN
94015 (86484)	**B**	P	CA	94029 (86740)	**P**	P	BK
94016 (86317)	**B**	P	CA	94030 (86746)	**B**	P	BK

94031 (86747)	B	P	BK	94033 (86731)	P P BK
94032 (86730)	B	P	CA	94034 (86200)	B P BK

NMV   NEWSPAPER VAN

Dia. NM501/2. Mark 1. Standard GUVs modified as newspaper vans. vd or vd*. ETH 3 (3X*). For lot numbers see old number series. Formerly NLV. Not now used for News traffic.

94050 (93771) d*	B	P	HQ	94062 (93803) d*	B	P	HQ
94051 (93708) e	B	P	HQ	94068 (93424) d*	B	P	HQ
94052 (93709) e	B	P	HQ	94071 (93544) d*	B	P	HQ
94058 (93530) d*	B	P	HQ	94077 (93862) d*	B	P	CA
94061 (93763) d*	B	P	HQ				

NAA   PROPELLING CONTROL VEHICLE

Dia. NA508. (PCV) Mark 1. Class 307 driving trailers converted for use in propelling parcels trains out of termini. Fitted with roller shutter doors. Equipment fitted for communication between cab of PCV and locomotive. a.

94300 (75114)	RX	P	EN	94301 (75102)	RX	P	EN

NOX (GUV)   GENERAL UTILITY VAN (100 MPH ETH WIRED)

Dia. NO513. xy. ETH 0 (0X*). Commonwealth bogies. For lot Nos. see GUV section. Add 2 t to weight. 95181 also caried 95361 and 95190 – 2 also carried 95393/91/92.

95100 (93668)	B	P	BK	95125 (93143)	B	P	BK
95101 (93142)	B	P	BK	95126 (93692)	B	P	BK
95102 (93762)	B	P	CA	95127 (93323)	P	P	BK
95103 (93956)	B	P	CA	95128 (93764)	P	P	CA
95104 (93942)	B	P	CA	95129 (93347)	P	P	EN
95105 (93126)	RX	P	CA	95130 (93263)	P	P	BK
95106 (93353)	RX	P	CA	95131 (93860)	B	P	BK
95107 (93576)	RX	P	CA	95132 (93607)	P	P	CA
95108 (93600)	B	P	BK	95133 (93604)	RX	P	EN
95109 (93269)	B	P	CA	95134 (93462)	RX	P	BK
95110 (93393)	B	P	BK	95135 (93249)	P	P	CA
95111 (93578)	B	P	EN	95136 (93396)	RX	P	BK
95112 (93673)	B	P	BK	95137 (93160)	RX	P	CA
95113 (93235)	B	P	CA	95138 (93212)	RX	P	BK
95114 (93081)	B	P	BK	95139 (93172)	P	P	CA
95115 (93174)	B	P	BK	95140 (93571)	P	P	BK
95116 (93426)	B	P	CA	95141 (93362)	RX	P	BK
95117 (93534)	B	P	BK	95142 (93844)	RX	P	CA
95118 (93675)	B	P	CA	95143 (93485)	B	P	CA
95119 (93167)	RX	P	CA	95144 (93165)	P	P	EN
95120 (93468)	P	P	BK	95145 (93293)	RX	P	EN
95121 (93518)	B	P	BK	95146 (93648)	RX	P	EN
95122 (93864)	B	P	CA	95147 (93091)	P	P	CA
95123 (93376)	B	P	CA	95148 (93416)	RX	P	BK
95124 (93836)	P	P	CA	95149 (93265)	P	P	BK

95150	(93560)	**P**	P	CA	95175	(93521)	**RX**	P	CA

95150 (93560)	**P**	P	CA
95151 (93606)	**RX**	P	EN
95152 (93969)	**P**	P	BK
95153 (93798)	**P**	P	EN
95154 (93897)	**P**	P	BK
95155 (93820)	**P**	P	BK
95156 (93160)	**RX**	P	BK
95157 (93523)	**B**	P	CA
95158 (93499)	**RX**	P	EN
95159 (93084)	**RX**	P	BK
95160 (93581)	**RX**	P	CA
95161 (93205)	**RX**	P	EN
95162 (93122)	**P**	P	EN
95163 (93407)	**RX**	P	CA
95164 (93104)	**RX**	P	CA
95165 (93262)	**RX**	P	EN
95166 (93112)	**RX**	P	BK
95167 (93255)	**RX**	P	BK
95168 (93914)	**RX**	P	CA
95169 (93277)	**RX**	P	BK
95170 (93395)	**P**	P	CA
95171 (93110)	**RX**	P	BK
95172 (93429)	**RX**	P	BK
95173 (94076)	**RX**	P	BK
95174 (93852)	**RX**	P	CA
95175 (93521)	**RX**	P	CA
95176 (93210)	**RX**	P	BK
95177 (93411)	**RX**	P	CA
95178 ( )			
95179 ( )			
95180 ( )			
95181 (94078) d*	**B**	P	BK
95182 ( )			
95183 ( )			
95184 ( )			
95185 ( )			
95186 ( )			
95187 ( )			
95188 ( )			
95189 ( )			
95190 (93278)	**P**	P	BK
95191 (93495)	**B**	P	BK
95192 (93643)		P	BK
95193 (93694)	**RX**	P	EN
95194 (93192)	**RX**	P	EN
95195 (93539)	**RX**	P	CA
95196 (93775)	**RX**	P	EN
95197 (93590)	**RX**	P	BK
95198 (93134)	**RX**	P	BK
95199 (93141)	**RX**	P	CA

NCX   NEWSPAPER VAN (100 mph)

Dia. NC501. BGs modified to carry newspapers. xe. ETH 3 (3X*). Commonwealth bogies. For lot Nos. see BG section. Add 2 t to weight. Not now used for News traffic. 95227/8/9/30 also carried 95310/32/29/21.

95200 (84019) *	**P**	P	BK	95216 (84542)		P	HQ
95201 (80875)	**P**	P	NC	95217 (84385)	**B**	P	EN
95202 (80667) *	**B**	P	BK	95218 (80675) *		P	BK
95204 (80947) *	**RX**	P	HT	95219 (80946)	**B**	P	BK
95205 (80620) *	**B**	P	BK	95220 (80717) *	**B**	P	BK
95206 (80561)	**P**	P	NC	95221 (84153) *		P	BK
95207 (80560) *	**RX**	P	HT	95222 (80774)	**RX**	P	HT
95208 (80660) *	**P**	P	BK	95223 (80933) *	**P**	P	NC
95209 (84047)	**RX**	P	HT	95227 (81292)	**RX**	P	NC
95210 (80731)	**RX**	P	BK	95228 (81014)	**RX**	P	NC
95211 (80949)	**P**	P	NC	95229 (81381)	**RX**	P	BK
95212 (84179)	**B**	P	BK	95230 (80525)	**RX**	P	NC
95214 (84360)	**B**	P	BK				

NCV   NEWSPAPER VAN (100 mph)

Dia. NC501. ve. Commonwealth bogies. For lot Nos. see BG section. Add 2t to weight. Not now used for News traffic.

95300 (80689)	**B**	P	HQ	95312 (80503)	**B**	P	HQ
95303 (80614)	**B**	P	HQ				

NOV (GUV)   GENERAL UTILITY VAN (100 MPH ETH WIRED)

Dia. NO513. xy. ETH 0X*. Commonwealth (B4*) bogies. For lot Nos. see GUV section. Add 2 t to weight for Commonwealth bogies.

95350 (93624)	**B**	P	CA	95373 (93258)	**RX**	P	CA
95351 (93596)	**RX**	P	CA	95374 (93367)	**RX**	P	CA
95352 (93727)	**B**	P	CA	95375 (	)		
95353 (93514)	**B**	P	CA	95376 (	)		
95354 (93725)	**B**	P	HQ	95377 (	)		
95355 (93375)	**B**	P	HQ	95378 (	)		
95356 (93478)	**B**	P	HQ	95379 (	)		
95357 (93508)	**B**	P	HQ	95380 (	)		
95358 (93195)	**B**	P	HQ	95381 (	)		
95359 (93854)	**B**	P	HQ	95382 (	)		
95360 (93207)	**B**	P	HQ	95383 (	)		
95362 (93563)	**RX**	P	CA	95384 (	)		
95363 (93345)	**RX**	P	CA	95385 (	)		
95364 (93715)	**B**	P	CA	95386 (	)		
95365 (93857)	**B**	P	HQ	95387 (	)		
95366 (93251)	**B**	P	HQ	95388 (	)		
95367 (93529)	**B**	P	HQ	95389 (	)		
95368 (93656)	**RX**	P	CA	95390 (	)		
95369 (93236)	**B**	P	CA	95391 (	)		
95370 (93710)	**B**	P	HQ	95392 (	)		
95371 (93713)	**P**	P	CA	95393 (	)		
95372 (93728)	**B**	P	HQ	95394 (	)		

NRX   BAA CONTAINER VAN (100 mph)

Dia. NR503. Modified for carriage of British Airports Authority containers with roller shutter doors and roller floors and gangways removed. xe. ETH 3 (3X*). Commonwealth bogies. For lot Nos. see BG section. Add 2 t to weight. Also carried 95203/13 respectively.

95400 (80621)	**RX**	P	DY	95410 (80826)	**RX**	P	DY

NX/NP (GUV)   MOTORAIL VAN

Mark 1. Dia. NX501. Renumbered from 93XXX series. For details and lot numbers see 93XXX series. ETH 0 (0X*).

96100 (93734) a*B5 **I** IWCX WB			96135 (93755) a C **I** ICCX EC	
96101 (93741) a*B5 **I** IWCX WB			96136 (93735) a C **I** IXXH HQ	
96103 (93744) a*B5 **I** IWCX WB			96137 (93748) a C **B** IXXH HQ	
96110 (93738) a*C **I** IWCX WB			96138 (93749) a C **I** IXXH HQ	
96111 (93742) a*C **I** IWCX WB			96139 (93751) a C **I** ICCX EC	
96112 (93750) a*C **I** ICCX EC			96141 (93753) a C **B** IXXH HQ	
96130 (93736) a*C **I** IWCX WB			96150 (93097) x*B5 **I** IWCX WB	
96131 (93737) a*C **I** IWCX WB			96155 (93334) x*B5 **I** IWCX WB	
96132 (93754) a*C **I** ICCX EC			96156 (93337) x*B5 **I** IWCX WB	
96133 (93685) a C **I** IXXH HQ			96157 (93344) x*B5 **I** IWCX WB	
96134 (93691) a C **I** IXXH HQ			96162 (93647) a*C **I** IWCX WB	

96163 (93646) a*C	I	IWCX	WB		96178 (93782) a*C	I	IWCX	WB
96164 (93880) a*C	I	IWCX	WB		96179 (93910) a*C	I	IWCX	WB
96165 (93784) a*C	I	IWCX	WB		96181 (93875) a*C	I	IWCX	WB
96166 (93834) a*C	I	IWCX	WB		96182 (93944) a*C	I	ICCX	EC
96167 (93756) a*C	I	IWCX	WB		96185 (93083) x*C	I	IWCX	WB
96168 (93978) a*C	I	IWCX	WB		96186 (93087) x*C	I	ICCX	EC
96169 (93937) a*C	I	IWCX	WB		96187 (93168) x*C	I	ICCX	EC
96170 (93159) x*C	I	IWCX	WB		96188 (93320) x*C	I	IWCX	WB
96171 (93326) x*C	I	IWCX	WB		96189 (93447) x*C	I	IWCX	WB
96172 (93363) x*C	I	IWCX	WB		96190 (93448) x*C	I	IWCX	WB
96173 (93440) x*C	I	IWCX	WB		96191 (93665) x*C	I	IWCX	WB
96174 (93453) x*C	I	ICCX	EC		96192 (93669) x*C	I	IWCX	WB
96175 (93628) x*C	I	IWCX	WB		96193 (93874) x*C	I	IWCX	WB
96176 (93641) x*C	I	IWCX	WB		96194 (93949) x*C	I	IWCX	WB
96177 (93980) x*C	I	IWCX	WB		96195 (93958) x*C	I	IWCX	WB

Mark 1. Dia. NP503. Vehicles modified with concertina end doors.

96210 – 8 were renumbered from 96159/04/61/52/54/58/60/53/51

96210 (93355) x*B5 I	IWCX	EN		96220 ( )	I	ICCX	EC
96211 (93745) a*B5 I	IWCX	EN		96221 ( )	I	ICCX	EC
96212 (93443) x*B5 I	IWCX	EN		96222 ( )	I	IWCX	WB
96213 (93324) x*B5 I	IWCX	EN		96223 ( )	I	IWCX	WB
96214 (93154) x*B5 I	IWCX	EN		96224 ( )	I	IWCX	WB
96215 (93351) x*B5 I	IWCX	EN		96225 ( )	I	IWCX	WB
96216 (93385) x*B5 I	IWCX	EN		96226 ( )	I	IWCX	WB
96217 (93327) x*B5 I	IWCX	EN		96227 ( )	I	IWCX	WB
96218 (93286) x*B5 I	ICCX	EN		96228 ( )	I	IWCX	WB
96219 ( )				96229 ( )	I	IWCX	WB

NG (MRCF)   MOTORAIL CAR FLAT

Dia. NG503a. These vehicles have been renumbered from wagons and are used for loading purposes.

96450	IWCX	WB		96452	IWCX	CL
96451	IWCX	WB		96453	IWCX	HQ

NY   EXHIBITION VAN

Various interiors. Converted from various vehicle types. Electric heating from shore supply. In some cases new lot numbers were issued for conversions, but not always. Non-standard livery – varies according to job being undertaken.

Lot 30842 Swindon 1972 – 3. Dia. NY503. Converted from BSK to Lot No. 30156 Wolverton 1955.

99621 (34697) x	BR1	0	ICHH	HQ	Exhibition Coach.
99625 (34693) x	Mk4	0	ICHH	HQ	Generator Van.

Converted Salisbury 1981 from RB to Lot No. 30636 Pressed Steel 1962. Dia NY523/4 respectively.

99645 (1765) v	C	0	ICHH	BN	Club Car.
99646 (1766) v	C	0	ICHH	BN	Club Car.

STORED COACHES

A number of coaches are stored and shown as allocated to 'HQ' in the main lists. Their last known locations are shown in this list, which may include certain coaches which are not necessarily stored, but have been given an 'HQ' allocation by BR for other reasons.

1646	Old Oak Common CARMD	5745	MoD Kineton
1649	Bounds Green T&RSMD	5750	MoD Kineton
1673	Bounds Green T&RSMD	5755	BRML Wolverton
1959	BRML Glasgow	5763	MoD Kineton
1966	Ferme Park	5768	MoD Kineton
1971	Ferme Park	5773	MoD Kineton
1972	Ferme Park	5775	MoD Kineton
1984	Ferme Park	5785	MoD Kineton
3192	Carlisle Upperby T&RSMD	5786	Wembley Intercity CARMD
3202	Carlisle Upperby T&RSMD	5787	MoD Kineton
3223	MoD Long Marston	5795	MoD Kineton
3225	MoD Kineton	5810	MoD Kineton
3226	MoD Kineton	5815	BRML Wolverton
3245	MoD Kineton	5823	Kirkdale EMUD
3258	MoD Kineton	5832	Crewe South Yard
3268	MoD Kineton	5834	MoD Kineton
4946	Edge Hill CARMD	5838	MoD Kineton
4979	Edge Hill CARMD	5844	Wembley Intercity CARMD
4996	Edge Hill CARMD	5849	MoD Kineton
5617	MoD Long Marston	5870	MoD Kineton
5621	MoD Kineton	5872	MoD Kineton
5633	MoD Long Marston	5879	MoD Long Marston
5637	MoD Long Marston	5881	MoD Kineton
5639	Crewe South Yard	5884	ABB Crewe
5640	BRML Wolverton	5886	MoD Kineton
5648	MoD Long Marston	5901	MoD Kineton
5659	Polmadie CARMD	5904	BRML Glasgow
5660	MoD Long Marston	6310	St. Philips Marsh T&RSMD
5661	MoD Kineton	6350	Old Oak Common CARMD
5663	MoD Kineton	6380	North Pole International Depot
5674	MoD Kineton	6381	North Pole International Depot
5675	MoD Long Marston	6382	North Pole International Depot
5687	MoD Kineton	6383	North Pole International Depot
5694	MoD Kineton	6605	MoD Kineton
5699	BRML Wolverton	6608	MoD Kineton
5701	MoD Kineton	6609	MoD Kineton
5706	Kirkdale EMUD	6614	MoD Kineton
5716	MoD Kineton	6619	MoD Kineton
5717	MoD Kineton	9482	BRML Wolverton
5718	MoD Kineton	10550	MoD Kineton
5729	MoD Long Marston	10567	Crewe South Yard
5731	MoD Kineton	10570	MoD Kineton
5738	MoD Kineton	10577	Wembley Intercity CARMD
5742	Kirkdale EMUD	10578	MoD Kineton

10579 MoD Kineton	80222 Gloucester New Yard
10581 Wembley Intercity CARMD	80223 Derby Etches Park T&RSMD
10591 MoD Kineton	80224 Derby Etches Park T&RSMD
10592 MoD Kineton	80225 Clapham Junction
10595 MoD Kineton	80305 Cambridge T&RSMD
10599 MoD Kineton	80310 Cambridge T&RSMD
10603 MoD Kineton	80312 Cambridge T&RSMD
10606 MoD Kineton	80313 Derby RTC
10609 MoD Kineton	80315 Cambridge T&RSMD
10656 MoD Kineton	80730 Ilford EMUD
10661 Rosyth Dockyard	80889 Doncaster West Yard
10662 Wembley Intercity CARMD	84044 Eastleigh
10665 Ferme Park	84197 Sheffield
10670 MoD Kineton	84234 BRML Glasgow
10673 Willesden New Yard	84399 Eastleigh
10678 MoD Kineton	84419 Doncaster West Yard
10679 MoD Kineton	84445 Doncaster West Yard
10684 MoD Kineton	84456 ABB Crewe
10700 MoD Kineton	84519 Crewe South Yard
10704 BRML Wolverton	84531 Eastleigh
10705 Wembley Intercity CARMD	92130 Gloucester New Yard
10707 Bounds Green T&RSMD	92133 Gloucester New Yard
10713 ABB Crewe	92158 Gloucester New Yard
10715 ABB Crewe	92161 BRML Glasgow
10720 MoD Kineton	92162 BRML Glasgow
10728 BRML Wolverton	92166 Gloucester New Yard
13581 Wembley Intercity CARMD	92168 BRML Glasgow
13583 Wembley Intercity CARMD	92172 Gloucester New Yard
13592 BRML Wolverton	92177 Doncaster West Yard
13593 ABB Derby	92184 Gloucester New Yard
13595 BRML Wolverton	92186 ABB Derby
13596 BRML Wolverton	92187 Doncaster West Yard
13601 BRML Wolverton	92193 ABB Derby
13603 BRML Wolverton	92198 Doncaster West Yard
17148 MoD Kineton	92203 Crewe South Yard
17155 MoD Kineton	92205 BRML Glasgow
17158 BRML Wolverton	92396 Eastleigh
17161 Bounds Green T&RSMD	92602 Gloucester New Yard
17163 MoD Kineton	92663 Doncaster West Yard
17171 Carlisle Upperby T&RSMD	92704 BRML Glasgow
80204 Derby Etches Park T&RSMD	92875 ABB Crewe
80206 Derby Etches Park T&RSMD	92971 Doncaster West Yard
80207 Gloucester New Yard	92989 Wembley Intercity CARMD
80208 Norwich Crown Point	92990 Carlisle Upperby T&RSMD
80211 Gloucester New Yard	93131 Gloucester New Yard
80213 Gloucester New Yard	93135 Crewe South Yard
80214 Gloucester New Yard	93267 Cambridge T&RSMD
80217 Gloucester New Yard	93273 Gloucester New Yard
80218 Gloucester New Yard	93394 Crewe South Yard
80219 Carlisle	93446 Gloucester New Yard
80220 Gloucester New Yard	93511 Derby Etches Park T&RSMD
80221 Gloucester New Yard	93515 ABB Crewe

93525 Derby Etches Park T&RSMD
93541 Gloucester New Yard
93562 Crewe South Yard
93585 Derby Etches Park T&RSMD
93660 Gloucester New Yard
93701 Cambridge T&RSMD
93714 Oxford CS
93717 Cambridge T&RSMD
93720 Derby Etches Park T&RSMD
93823 Gloucester New Yard
93830 Cambridge T&RSMD
93847 Gloucester
93849 Derby Etches Park T&RSMD
93859 ABB Crewe
93881 Derby Etches Park T&RSMD
93886 Gloucester New Yard
93905 Gloucester New Yard
93952 Gloucester New Yard
93955 ABB Crewe
93962 Gloucester New Yard
93973 ABB Crewe
93979 Gloucester New Yard
94050
94051 Cambridge T&RSMD
94052 Cambridge T&RSMD
94058 Cambridge T&RSMD
94061 Gloucester New Yard
94062 Crewe South Yard
94068 Derby Etches Park T&RSMD
94071 Crewe South Yard
95312 Gloucester New Yard
95354 Derby Etches Park T&RSMD
95356 Derby Etches Park T&RSMD
95357 Derby Etches Park T&RSMD
95358 Gloucester New Yard
95359 Derby Etches Park T&RSMD
95360 Gloucester New Yard
95365 Derby Etches Park T&RSMD
95366 Gloucester
95367 Gloucester New Yard
95370 Eastleigh
95372 Gloucester New Yard
96133 Craigentinny T&RSMD
96134 Craigentinny T&RSMD
96136 Craigentinny T&RSMD
96137 Craigentinny T&RSMD
96138 Craigentinny T&RSMD
96141 Craigentinny T&RSMD
99621 Ferme Park
99625 Ferme Park

PRIVATELY OWNED COACHES

This list comprises privately owned coaches which are "plated" to run on BR
The PO number is carried on a yellow plate affixed to the solebar. Coaches have
to be passed by BR each year and this is denoted by a white equilateral triangle
painted on the solebar with the year painted in black. In the following list, the
original number is shown in column 3. It should be noted that other numbers
may also have been carried. It should be noted that there are other vehicles,
not at present registered for running, which also have BR private-owner numbers
allocated.

Prefix	No.	Old No.	Type	Base	Comments
RRT	99020	5476	BR Mark 2B TSO	CS	Ridings Railtours Set
RRT	99021	5533	BR Mark 2C TSO	CS	Ridings Railtours Set
RRT	99022	5574	BR Mark 2C TSO	CS	Ridings Railtours Set
RRT	99023	5585	BR Mark 2C TSO	CS	Ridings Railtours Set
RRT	99024	5595	BR Mark 2C TSO	CS	Ridings Railtours Set
PWDS	99030	34666	BR Mark 1 BSK	CS	5407 support coach
BLS	99035	35322	BR Mark 1 BSK	DI	70000 support coach
MRC	99040	21232	BR Mark 1 BCK	SK	80080 support coach
MRC	99041	35476	BR Mark 1 BSK	SK	46203 support coach
FSS	99050	1	GER insp'n saloon	SO	Sir W.H.Mcalpine's sal.
SHRC	99052	484	WCJS dining saloon	PB	'Queen of Scots' train
RFM	99053	9004	GWR first saloon	CS	Railfilms Ltd saloon
SPG	99070	35123	BR Mark 1 BSK	BT	44871 support coach
SNG	99080	21096	BR Mark 1 BCK	CS	4498 support coach
RPR	99120	21236	BR Mark 1 BCK	KR	SVR locos supp. coach
WDS	99121	3105	BR Mark 1 FO	CS	Carnforth maroon set
FSS	99122	3106	BR Mark 1 FO	BN	BN91 set
FSS	99123	3109	BR Mark 1 FO	BN	BN91 set
FSS	99124	3110	BR Mark 1 FO	BN	BN91 set
WDS	99127	3117	BR Mark 1 FO	CS	Carnforth maroon set
FSS	99129	21272	BR Mark 1 BCK	BN	BN91 set
GSWR	99131	1531	LNER prototype first	MH	'Royal Scotsman' train
FSS	99132	1861	BR Mark 1 RMB	BN	BN91 set
CHEL	99140	796	GWR third	DI	Reb. as dynamometer car
CHEL	99141	14041	BR Mark 2 BFK	DI	71000 support coach
GWS	99180	35333	BR Mark 1 BSK	DI	6024 support coach
FSS	99190	3131	BR Mark 1 FO	BN	On lease from BR. BN91
FSS	99191	3132	BR Mark 1 FO	BN	On lease from BR. BN91
FSS	99192	3133	BR Mark 1 FO	BN	On lease from BR. BN91
FSS	99193	4860	BR Mark 1 TSO	BN	On lease from BR. BN91
FSS	99194	5032	BR Mark 1 TSO	BN	On lease from BR. BN91
FSS	99195	5035	BR Mark 1 TSO	BN	On lease from BR. BN91
SVR	99241	35449	BR Mark 1 BSK	SO	34027 support coach
SVR	99242	35467	BR Mark 1 BSK	KR	SVR locos supp. coach
WDS	99302	13323	BR Mark 1 FK	CS	Carnforth maroon set
WDS	99303	13317	BR Mark 1 FK	CS	Carnforth maroon set
WDS	99304	21256	BR Mark 1 BCK	CS	Carnforth maroon set
WDS	99311	1882	BR Mark 1 RMB	CS	Carnforth maroon set
WDS	99312	35463	BR Mark 1 BSK	CS	48151 support coach

DRC	99313	35451	BR Mark 1 BSK	HH	45596 support coach
WDS	99314	25729	BR Mark 1 SK	CS	Carnforth maroon set
WDS	99315	25955	BR Mark 1 SK	CS	Carnforth maroon set
WDS	99316	13321	BR Mark 1 FK	CS	Carnforth maroon set
WDS	99317	3766	BR Mark 1 TSO	CS	Carnforth maroon set
WDS	99318	4912	BR Mark 1 TSO	CS	Carnforth maroon set
WDS	99319	14168	BR Mark 2D BFK	CS	Carnforth maroon set
WDS	99321	5299	BR Mark 2A TSO	CS	Carnforth maroon set
WDS	99322	5600	BR Mark 2C TSO	CS	Carnforth maroon set
WDS	99323	5704	BR Mark 2D TSO	CS	Carnforth maroon set
WDS	99324	5714	BR Mark 2D TSO	CS	Carnforth maroon set
WDS	99325	5727	BR Mark 2D TSO	CS	Carnforth maroon set
FSS	99357	3112	BR Mark 1 FO	BN	BN91 set
WDS	99358	3128	BR Mark 1 FO	CS	Carnforth maroon set
JBC	99405	35486	BR Mark 1 BSK	MK	60009 support coach
HLPG	99421	14021	BR Mark 1 BFK	SO	777 Support Coach
HLPG	99423	4828	BR Mark 1 SO	CP	
HLPG	99424	4823	BR Mark 1 SO	CP	
HLPG	99425	4822	BR Mark 1 SO	CP	
HLPG	99427	35204	BR Mark 1 BSK	SO	777 support coach
GWS	99512	34671	BR Mark 1 BSK	DI	5029 support coach
VSOE	99530	301	Pullman parlour first	SL	'PERSEUS'
VSOE	99531	302	Pullman parlour first	SL	'PHOENIX'
VSOE	99532	308	Pullman parlour first	SL	'CYGNUS'
VSOE	99533	70741	LNER BGP	SL	'BAGGAGE CAR No. 7'
VSOE	99534	245	Pullman kitchen first	SL	'IBIS'
VSOE	99535	213	Pullman Brake First	SL	'MINERVA'
VSOE	99536	254	Pullman parlour first	SL	'ZENA'
VSOE	99537	280	Pullman kitchen first	SL	'AUDREY'
VSOE	99538	34991	BR Mark 1 BSK	SL	'BAGGAGE CAR No. 9'
VSOE	99539	255	Pullman kitchen first	SL	'IONE'
VSOE	99540	3069	BR Mark 1 FO	SL	'SALOON CAR No. 1'
VSOE	99541	243	Pullman kitchen first	SL	'LUCILLE'
VSOE	99542		Ferry van 889202	SL	'BAGGAGE CAR No. 8'
VSOE	99543	284	Pullman kitchen first	SL	'VERA'
MANC	99670	546	Pullman parlour first	CS	'CITY OF MANCHESTER'*
MANC	99671	548	Pullman parlour first	CS	'ELIZABETHAN'*
MANC	99672	549	Pullman parlour first	CS	'PRINCE RUPERT'*
MANC	99673	550	Pullman parlour first	CS	'GOLDEN ARROW'*
MANC	99674	551	Pullman parlour first	CS	'CALEDONIAN'*
MANC	99675	552	Pullman parlour first	CS	'SOUTHERN BELLE'*
MANC	99676	553	Pullman parlour first	CS	'KING ARTHUR'*
MANC	99677	586	Pullman brake first	CS	'TALISMAN'*
MANC	99678	504	Pullman kitchen first	CS	'THE WHITE ROSE'*
MANC	99679	506	Pullman kitchen first	CS	'THE RED ROSE'*
MANC	99680	14102	BR Mark 2A BFK	CS	'ATTENDANT'S CAR'*
TRTS	99710	25767	BR Mark 1 SK	CS	Pilkington K glass set
TRTS	99712	25893	BR Mark 1 SK	CS	Pilkington K glass set
TRTS	99713	26013	BR Mark 1 SK	CS	Pilkington K glass set
TRTS	99714	16187	BR Mark 1 CK	CS	Pilkington K glass set
TRTS	99716	25808	BR Mark 1 SK	CS	Pilkington K glass set
TRTS	99717	25837	BR Mark 1 SK	CS	Pilkington K glass set

TRTS	99718	25862	BR Mark 1 SK	CS	Pilkington K glass set
TRTS	99719	16191	BR Mark 1 CK	CS	Pilkington K glass set
TRTS	99720	35461	BR Mark 1 BSK	CS	Pilkington K glass set
TRTS	99721	25806	BR Mark 1 SK	CS	Pilkington K glass set
TRTS	99722	25756	BR Mark 1 SK	CS	Pilkington K glass set
TRTS	99723	35459	BR Mark 1 BSK	CS	Pilkington K glass set
NELP	99760	34557	BR Mark 1 BSK	WT	60532 support coach
SCR	99818	1730	BR Mark 1 RB	BT	SRPS set
SCR	99820	4871	BR Mark 1 TSO	BT	SRPS set
SCR	99821	9227	BR Mark 1 BSO	BT	SRPS set
SCR	99822	1859	BR Mark 1 RMB	BT	SRPS set
SCR	99823	4832	BR Mark 1 TSO	BT	SRPS set
SCR	99824	4831	BR Mark 1 TSO	BT	SRPS set
SCR	99825	13228	BR Mark 1 FK	BT	SRPS set
SCR	99826	13229	BR Mark 1 FK	BT	SRPS set
SCR	99827	3096	BR Mark 1 FO	BT	SRPS set
SCR	99828	13230	BR Mark 1 FK	BT	SRPS set
SCR	99829	4856	BR Mark 1 TSO	BT	SRPS set
SCR	99830	5028	BR Mark 1 TSO	BT	SRPS set
SHRC	99880	5159	LNWR dining saloon	PB	'Queen of Scots' train
SHRC	99881	807	GNR family saloon	PB	'Queen of Scots' train
SHRC	99886	35407	BR Mark 1 BSK	PB	'Queen of Scots' train
GSWR	99887	2127	BR Mark 1 SLF	MH	'Royal Scotsman' train
NRMY	99951	10656	SR DMBTK (EMU)	BI	National collection
NRMY	99952	12123	SR DTCK (2 Bil)	BI	National collection
NRMY	99953	35468	BR Mark 1 BSK	YK	NRM Locos supp. coach
GSWR	99961	324	Pullman parlour first	MH	'Royal Scotsman' train
GSWR	99962	329	Pullman parlour first	MH	'Royal Scotsman' train
GSWR	99963	331	Pullman parlour first	MH	'Royal Scotsman' train
GSWR	99964	313	Pullman kitchen first	MH	'Royal Scotsman' train
GSWR	99965	319	Pullman kitchen first	MH	'Royal Scotsman' train
GSWR	99966	34525	BR Mark 1 BSK	MH	'Royal Scotsman' train
GSWR	99967	317	Pullman kitchen first	MH	'Royal Scotsman' train
EAB	99990	14025	BR Mark 1 BFK	CS	46441 support coach

*Manchester Pullman set.

Note: RPR 99120 is on loan to the Eastleigh RPS for use as 828 support coach.

We would like to thank Mr. P. Hall for information on preserved coaches.

OWNER PREFIX CODES

BLS	Britannia locomotive Society.
CHEL	Chell Instruments, Tudor House, Grammar School Road, North Walsham Norfolk.
DRC	Bahamas Locomotive Society.
EAB	Mrs. E. Beet, Claughton, 63 Marine Drive, Hest Bank, Lancaster.
FSS	Flying Scotsman Railways, 104 Birmingham Rd, Lichfield, Staffs.
GSWR	Great Scottish & Western Railway Co. Ltd., c/o L & R leisure plc, 46a Constitution Street, Leith, Edinburgh, EH6 6RS.
GWS	Great Western Society, Didcot Railway Centre, Didcot, Oxon.
HLPG	Humberside Locomotive Preservation Group, 19 Wilson Street,

	Anlaby, Hull, HU10 7AN.
JBC	J.B. Cameron, Balbuthie Farm, Kilconquhar, Fife.
MANC	Manchester Pullman Co. Ltd., 2nd Floor, Yorkshire Bank Chambers, St. James Street, Accrington, Lancs, BB5 1LY.
MRC	B.P. Ewart, The Holmestead, Yeldersley Lane, Bradley, Ashbourne, DE6 1PJ.
NELP	North Eastern Loco. Preservation Group, 21 Front Street, Daisy Hill, Sacriston, Durham, DH7 6BL.
NRMY	National Railway Museum, Leeman Road, York.
PWDS	P.W.D. Smith, Orchard Lea, Lady Lane, Mobberly, Knutsford.
RFM	Railfilms Ltd., Mr. N. Dobson, 26 Regent Street, Altrincham, Cheshire, WA14 1RP.
RPR	RPR Railway Coach Association, c/o Charles Paget, 11 Western Close, Dorridge, Solihull, B93 8BL.
RRT	Ridings Railtours, c/o Dr. Adrian Morgan, P.O Box 18, Ripon, N. Yorks, HG4 2ER.
SCR	Scottish Railway Preservation Society, Macland, Maddiston Road, Brightons, Falkirk, FK2 0JP.
SHRC	Scottish Highland Railway Co. Ltd., Mills Road, Aylesford, Maidstone, Kent, ME20 7NW.
SNG	A4 Locomotive Society, Steamtown Railway Museum, Warton, Carnforth, Lancs, LA5 9HX.
SPG	44871 Sovereign Preservation Group, 15 Montrose Gardens, Milngavie, Glasgow, G62 8NQ.
SVR	Severn Valley Railway, Bewdley Station, Bewdley, Worcestershire.
TRTS	Train Tours (Charter Rail) Ltd., 53 Rutland Street, St. Helens, Merseyside, WA10 2BY.
VSOE	Venice Simplon Orient Express Ltd., 1 Hanover Square, London, W1R 9RD.
WDS	W.D. Smith, Steamtown Railway Museum, Carnforth, Lancs, LA5 9HX.

PRIVATE OWNER BASE CODES

Official BR codes have now been issued for private owner bases.

BT	Bo'Ness & Kinneil Railway, Bo'Ness Station, West Lothian.
BN	BR Bounds Green T&RSMD, London.
IS	BR Inverness CS, Highland Region.
MH	Great Scottish & Western Railway Company, BR Millerhill Royal Scotsman Siding, Edinburgh, Lothian.
DI	Great Western Society, Didcot Railway Centre, Didcot, Oxon.
HH	Keighley & Worth Valley Railway, Ingrow, Keighley, West Yorks.
WT	ICI Wilton, Cleveland.
MK	Markinch Goods Shed, Markinch, Fife.
SK	Midland Railway Centre, Swanwick Junction, Derbyshire.
YK	National Railway Museum, Leeman Road, York, North Yorkshire.
PB	Nene Valley Railway, Wansford Station, Peterborough, Cambs.
SL	BR Stewarts Lane T&RSMD, London.
CS	Steamtown Railway Centre, Carnforth, Lancashire.
SO	Southall Railway Centre, Southall Depot, Southall, Gtr. London
KR	Severn Valley Railway, Kidderminster, Worcestershire.

SECTOR CODES

GPSG	European Passenger Services. DVT.
GPSM	European Passenger Services. Barrier vehicles.
IANR	Intercity Anglia. Refurbished or re-trimmed.
IANX	Intercity Anglia.
ICCE	Intercity Cross Country. EC HST trailers.
ICCG	Intercity Cross Country. Barrier vehicles.
ICCL	Intercity Cross Country. On loan.
ICCR	Intercity Cross Country. Refurbished or re-trimmed.
ICCT	Intercity Cross Country. LA and NL HST trailers.
ICCX	Intercity Cross Country.
ICHD	Intercity Charter and Special Services. Air braked not refurbished.
ICHH	Intercity Charter and Special Services. Held for re-use.
ICHL	Intercity Charter and Special Services. For loan.
ICHS	Intercity Charter and Special Services. Steam specials.
ICHV	Intercity Charter and Special Services. VIP.
ICHX	Intercity Charter and Special Services.
IECD	Intercity East Coast Main Line. HST trailers.
IECG	Intercity East Coast Main Line. Barrier vehicles.
IECX	Intercity East Coast Main Line.
ILAG	Intercity Barrier Vehicles Laira – Derby.
IMLG	Intercity Midland Main Line. Barrier vehicles.
IMLR	Intercity Midland Main Line. Refurbished.
IMLX	Intercity Midland Main Line.
IWCD	Intercity West Coast Main Line. Barrier vehicles.
IWCR	Intercity West Coast Main Line. Refurbished.
IWCX	Intercity West Coast Main Line.
IWRG	Intercity Great Western Main Line. Barrier vehicles.
IWRR	Intercity Great Western Main Line. Refurbished.
IWRX	Intercity Great Western Main Line.
IXXB	Intercity. Held for bogie re-use.
IXXC	Intercity. Secondary door modifications float.
IXXH	Intercity. Held for further use.
IXXT	Intercity. Awaiting transfer.
IXXZ	Intercity. Authorised for withdrawal.
NSSX	Network SouthEast. Solent and Sarum services.
NWXX	Network SouthEast. West of England/North Downs services.
NXXZ	Network SouthEast. Authorised for withdrawal.
P	Rail Express Systems.
RAIS	Regional Railways Scotrail. Inverness.
RAXX	Regional Railways Scotrail. Inverness. Stored.
RBHT	Regional Railways North East. Heaton.
RCLG	Regional Railways North West. Longsight.
RCLL	Regional Railways North West. Edge Hill.
RCXX	Regional Railways North West. Edge Hill. Stored.
RFXZ	Regional Railways Headquarters. Authorised for withrawal.